Advances in MATHEMATICAL ECONOMICS

Aims and Scope. The project is to publish *Advances in Mathematical Economics* once a year under the auspices of the Research Center for Mathematical Economics. It is designed to bring together those mathematicians who are seriously interested in obtaining new challenging stimuli from economic theories and those economists who are seeking effective mathematical tools for their research.

The scope of *Advances in Mathematical Economics* includes, but is not limited to, the following fields:

- Economic theories in various fields based on rigorous mathematical reasoning.
- Mathematical methods (e.g., analysis, algebra, geometry, probability) motivated by economic theories.
- Mathematical results of potential relevance to economic theory.
- Historical study of mathematical economics.

Authors are asked to develop their original results as fully as possible and also to give a clear-cut expository overview of the problem under discussion. Consequently, we will also invite articles which might be considered too long for publication in journals.

S. Kusuoka, A. Yamazaki (Eds.)

Advances in Mathematical Economics

Volume 10

Springer

Shigeo Kusuoka
Professor
Graduate School of Mathematical Sciences
University of Tokyo
3-8-1 Komaba, Meguro-ku
Tokyo, 153-0041 Japan

Akira Yamazaki
Professor
Department of Economics
Meisei University
Hino
Tokyo, 191-8506 Japan

ISBN 978-4-431-54697-9 Springer Tokyo Berlin Heidelberg New York

Printed on acid-free paper
Springer is a part of Springer Science+Business Media
springer.com

Softcover re-print of the Hardcover 1st edition 2007

Camera-ready copy prepared from the authors' LaTeXfiles.

SPIN: 12024780

Table of Contents

Research Articles

Adv. Math. Econ. 10, 1–29 (2007)

Advances in
MATHEMATICAL
ECONOMICS

Komlós type convergence for random variables and random sets with applications to minimization problems

C. Castaing[1] **and M. Saadoune**[2]

[1] Département de Mathématiques, Université Montpellier II, 34095 Montpellier Cedex 5, France
(e-mail: castaing.charles@numericable.fr)

[2] Département de Mathématiques, Université Ibnou Zohr, Lot. Addalha, B.P. 8106, Agadir, Maroc
(e-mail: msaadoune@hotmail.com)

Received: August 10, 2006
Revised: October 16, 2006

JEL classification: C61

Mathematics Subject Classification (2000): 49J40, 49J45, 46N10

Abstract. Let E be a separable super reflexive Banach space and let $(\Omega, \mathcal{F}, \mu)$ be a complete probability space. We state some Komlós type theorems in the space $\mathcal{L}^0_E(\Omega, \mathcal{F}, \mu)$ of E-valued random variables and a version of Komlós slice theorem in the space $\mathcal{L}^0_{cwk(E)}(\Omega, \mathcal{F}, \mu)$ of convex weakly compact random sets. Weak Komlós type theorems for some unbounded sequences in $\mathcal{L}^1_F(\Omega, \mathcal{F}, \mu)$ and $\mathcal{L}^1_{F'}[F](\Omega, \mathcal{F}, \mu)$ when F is a separable Banach space are also stated. A Fatou type lemma in Mathematical Economics and minimization problems on convex and closed in measure subsets of $\mathcal{L}^0_E(\Omega, \mathcal{F}, \mu)$ are presented. Further Minimization problems and Min–Max type results involving saddle-points and Young measures are also investigated.

Key words: Biting Lemma, Komlós convergence, minimization, Min-Max, saddle points, young measures

1. Introduction and preliminaries

Throughout E is a separable Banach space, $(\Omega, \mathcal{F}, \mu)$ is a complete probability space, $\mathcal{L}^0_E(\Omega, \mathcal{F}, \mu)$ is the space of all $\mathcal{F}$-measurable E-valued functions defined on Ω. Let $cwk(E)$ be the set of all nonempty convex weakly compact subsets of E. Let us denote by $\mathcal{L}^0_{cwk(E)}(\Omega, \mathcal{F}, \mu)$ the space of all scalarly $\mathcal{F}$-measurable $cwk(E)$-valued mapping defined on Ω (alias convex weakly compact random sets). Recall that a $cwk(E)$-valued mapping $X : \Omega \to cwk(E)$ is scalarly $\mathcal{F}$-measurable if the support functions $\delta^*(x^*, X(.))$ are $\mathcal{F}$-measurable for all $x^* \in E'$. Given a convex weakly compact random set X, we denote by $|X|$ the $\mathcal{F}$-measurable real valued function $|X| : \Omega \to |X(\omega)| := \sup\{|\delta^*(x^*, X(\omega))| : \|x^*\| \leq 1\}$. We refer to [18] for details concerning Convex Analysis and Measurable Multifunctions. Also we will use the following limiting notions. Let $(C_n)_{n\in\mathbf{N}\cup\{\infty\}}$ be a sequence of nonempty closed convex subsets of E, $(C_n)_{n\in\mathbf{N}}$ *Mosco converges* to C_∞ if the two following inclusions are satisfied:

$$C_\infty \subset s\text{-}li C_n := \{x \in E : ||x - x_n|| \to 0; x_n \in C_n\}$$
$$w\text{-}ls C_n := \left\{x \in E : x_{n_k} \to x \text{ weakly}; x_{n_k} \in C_{n_k}\right\} \subset C_\infty.$$

Given two nonempty subsets B and C in E, the *gap* between B and C is defined by

$$D(B, C) = \inf\{||x - y|| : x \in B, y \in C\}.$$

The *slice topology* τ_s on $cc(E)$ (nonempty closed convex subsets of E) is the weakest topology τ on $cc(E)$ such that for each nonempty bounded closed convex subset B of E, the function $C \mapsto D(B, C)$ is τ-continuous. $(C_n)_{n\in\mathbf{N}}$ *slice converges* to C_∞ if one has

$$\lim_{n\to\infty} D(B, C_n) = D(B, C_\infty)$$

for all nonempty bounded closed convex subset B of E. It is well-known that the slice convergence and the Mosco convergence coincide on $cc(E)$ when E is reflexive. We refer to [6] for the topologies on closed convex subsets in Banach spaces. If F is a reflexive Banach space, any bounded sequence (f_n) in the space $\mathcal{L}^1_F(\Omega, \mathcal{F}, \mu)$ has the *Mazur property*; namely, there exist a subsequence (g_m) of (f_n) and $f_\infty \in \mathcal{L}^1_F(\Omega, \mathcal{F}, \mu)$ and a sequence (h_n) of convex combinations of (g_m), i.e. $h_n \in co\{g_m : m \geq n\}$, $\forall n \in \mathbf{N}$, such that (h_n) converges a.e. to f_∞, with respect to the norm topology of F. Further if F is super reflexive, then any bounded sequence (f_n) in $\mathcal{L}^1_F(\Omega, \mathcal{F}, \mu)$ has the *Komlós property*, namely, there exist a subsequence $(f_{\alpha(n)})$ in $\mathcal{L}^1_F(\Omega, \mathcal{F}, \mu)$ and $f_\infty \in \mathcal{L}^1_F(\Omega, \mathcal{F}, \mu)$, such that

$$\lim_{n\to\infty} \frac{1}{n} \Sigma_{j=1}^n f_{\beta(j)} = f_\infty$$

a.e., with respect to the norm topology, for every subsequence $(f_{\beta(n)})$ of $(f_{\alpha(n)})$. Let $(f_n)_{n\in\mathbf{N}\cup\{\infty\}}$ in $\mathcal{L}^0_E(\Omega, \mathcal{F}, \mu)$, the notation $\mu\text{-}\lim_{n\to\infty} f_n = f_\infty$ means that (f_n) converges to f_∞ in measure. For more on Komlós theorem [20] in $L^1_F(\Omega, \mathcal{F}, \mu)$ where F is super reflexive (or B-convex) space, we refer to [7, 19].

In § 2 we present several versions of Komlós theorem [20]. The first ones concerns with a version of Komlós theorem in $\mathcal{L}^0_E(\Omega, \mathcal{F}, \mu)$ and a version of Komlós slice type theorem on the space $\mathcal{L}^0_{cwk(E)}(\Omega, \mathcal{F}, \mu)$ when E is a super reflexive. The second ones deals with *weak* Komlós type theorems for some *unbounded* sequences in $\mathcal{L}^1_F(\Omega, \mathcal{F}, \mu)$ and $\mathcal{L}^1_{F'}[F](\Omega, \mathcal{F}, \mu)$, respectively, here $\mathcal{L}^1_{F'}[F](\Omega, \mathcal{F}, \mu)$ denotes the space of all F'-valued functions $f : \Omega \to F'$ such that for all $x \in F$, the scalar function $\langle x, f\rangle$ is integrable and such that the function $|f| : \Omega \to \mathbf{R}$ given by $|f|(\omega) := ||f(\omega)||_{F'}$, $\omega \in \Omega$ is integrable, when F is a separable Banach space, here the notation *weak* Komlós means that the associated Césaro sums converges a.e to the functions under consideration, with respect to the weak topology of F and the weak* topology of F', respectively. Our main purpose is to introduce a new type of *tightness condition* for sequences in these spaces. Namely, a sequence (f_n) in $\mathcal{L}^1_F(\Omega, \mathcal{F}, \mu)$, is *Mazur tight* if it satisfies the condition (*): for every subsequence (f_{n_k}) of (f_n) there exists a sequence (r_n) in $\mathcal{L}^1_{\mathbf{R}}(\Omega, \mathcal{F}, \mu)$ with $r_n \in co\{||f_{n_i}(.)|| : i \geq n\}$ such that $\limsup_n r_n \in \mathcal{L}^1_{\mathbf{R}}(\Omega, \mathcal{F}, \mu)$ and similarly for (g_n) in $\mathcal{L}^1_{F'}[F](\Omega, \mathcal{F}, \mu)$. It is worthy to mention that the above Mazur tightness condition does not imply that (f_n) is bounded in $\mathcal{L}^1_F(\Omega, \mathcal{F}, \mu)$. Indeed, it suffices to consider the space $\mathcal{L}^1_{\mathbf{R}}(\Omega, \mathcal{F}, \mu)$ where $\Omega = [0, 1]$ endowed with the Lebesgue measure and f_n is given by $f_n(\omega) := n^2 1_{[0,1/n]}(\omega)$, $\omega \in \Omega$. Then, (f_n) is not bounded in $\mathcal{L}^1_{\mathbf{R}}(\Omega, \mathcal{F}, \mu)$ but satisfies (*), because it converges a.e. to 0. These considerations led to several tightness conditions in the study of Fatou lemma in Mathematical Economics [15, 16] and allow to give a new light in the problem under consideration. In § 3 a series of new applications illustrating the results obtained in the previous sections is given. We present a characterization of convex closed sets for the convergence in measure by means of Komlós theorem in $\mathcal{L}^0_E(\Omega, \mathcal{F}, \mu)$. In particular we show that for every convex and closed in measure subset $\mathcal{H}$ of $\mathcal{L}^0_E(\Omega, \mathcal{F}, \mu)$ and for every convex weakly compact random set Γ, the set $\mathcal{S}_\Gamma + \mathcal{H}$ where $\mathcal{S}_\Gamma$ is the set of all $\mathcal{F}$-measurable selections of Γ, is convex and closed in measure. A Fatou-type lemma in Mathematical Economics and some Minimization problems are also given. We provide also two min–max type results involving Young measures and saddle points for a class of convex integral functionals.

We refer to [7, 14, 19, 20, 23] for Komlós theorem for vector-valued random variables and to [4, 12, 17, 21] for random sets.

2. Komlós theorem for random variables and random sets with applications

The following version of Komlós theorem in $\mathcal{L}^0_E(\Omega, \mathcal{F}, \mu)$ is crucial for our purpose.

Theorem 2.1. *Let E be a separable super reflexive Banach space and let (f_n) be a sequence in $\mathcal{L}_E^0(\Omega, \mathcal{F}, \mu)$ satisfying the following condition: (*) for every subsequence (f_{n_k}) of (f_n) there exists a sequence (r_n) in $\mathcal{L}_{\mathbf{R}}^0(\Omega, \mathcal{F}, \mu)$ with $r_n \in co\{||f_{n_i}(.)|| : i \geq n\}$ such that $\limsup_n r_n(\omega)$ is finite for each ω in Ω. Then there exist a subsequence (g_n) of (f_n) and an E-valued random vector f such that*

$$\lim_n ||\frac{1}{n}\sum_{i=1}^{n} h_i(\omega) - f(\omega)|| = 0, \quad a.e.\ \omega \in \Omega,$$

for every subsequence (h_n) of (g_n). Furthermore, if we suppose in this condition $\limsup_n r_n \in \mathcal{L}_{\mathbf{R}}^1(\Omega, \mathcal{F}, \mu)$, then $f_\infty \in \mathcal{L}_E^1(\Omega, \mathcal{F}, \mu)$.

Proof. By hypothesis, there exists a sequence (r_n) of the form

$$r_n = \sum_{i \in I_n} \lambda_i^n ||f_{i+n}||$$

with $\lambda_i^n \geq 0$ and $\sum_{i \in I_n} \lambda_i^n = 1$ such that $\limsup_n r_n(\omega)$ is finite (equivalently $\sup_n r_n(\omega)$ is finite) for each ω in Ω. For $p \in \mathbf{N}$, define

$$A_p := \{\omega \in \Omega : \sup_n r_n(\omega) \leq p\}.$$

As $\lim_{p\to\infty} \mu(A_p) = 1$, we can choose $p_1 \in \mathbf{N}$ such that $\mu(A_{p_1}) > 1 - \frac{1}{2}$. By integrating we get

$$\sup_n \sum_{i \in I_n} \lambda_i^n \int_{A_{p_1}} ||f_{i+n}(\omega)|| d\mu \leq p_1$$

Hence, there exists a subsequence (f_n^1) such that $\sup_n \int_{A_{p_1}} ||f_n^1(\omega)|| d\mu \leq p_1$. In view of the Biting lemma (see e.g. [13], Theorem 6.1.4) there exists an increasing sequence (B_q^1) in $\mathcal{F}$ with $\lim_q \mu(B_q^1) = 1$ and a subsequence (g_n^1) of (f_n^1) such that (g_n^1) is uniformly integrable on each $A_{p_1} \cap B_q^1$. Now, by virtue of Komlós theorem [19], there exist a subsequence of (g_n^1) still denoted by (g_n^1), $f_\infty^1 \in \mathcal{L}_E^1(\Omega, \mathcal{F}, \mu)$ and $\varphi_\infty^1 \in \mathcal{L}_{\mathbf{R}}^1(\Omega, \mathcal{F}, \mu)$ such that the following hold

$$\lim_{n\to\infty} ||\frac{1}{n}\Sigma_{i=1}^n h_i(\omega) - f_\infty^1(\omega)|| = 0, \quad \text{a.e. } \omega \in A_{p_1}$$

and

$$\lim_{n\to\infty} \frac{1}{n}\Sigma_{i=1}^n ||h_i(\omega)|| = \varphi_\infty^1, \quad \text{a.e. } \omega \in A_{p_1}$$

for every subsequence (h_n) of (g_n^1). As (g_n^1) is uniformly integrable on each $B_q^1 \cap A_{p_1}$, by Lebesgue–Vitali theorem, we get

$$\forall q, \forall A \in \mathcal{F}, \quad \lim_{n\to\infty} \int_{A\cap A_{p_1}\cap B_q^1} \frac{1}{n}\Sigma_{i=1}^n ||h_i|| d\mu = \int_{A\cap A_{p_1}\cap B_q^1} \varphi_\infty^1 d\mu$$

for every subsequence (h_n) of (g_n^1). This is equivalent to

$$\forall q, \forall A \in \mathcal{F}, \quad \lim_{n\to\infty} \int_{A\cap A_{p_1}\cap B_q^1} ||g_n^1|| d\mu = \int_{A\cap A_{p_1}\cap B_q^1} \varphi_\infty^1 d\mu.$$

Next, applying condition (*) to (g_n^1) instead of (f_n) and using again the preceding arguments we find a measurable set A_{p_2} with $\mu(A_{p_2}) > 1 - \frac{1}{3}$, an increasing sequence (B_q^2) in $\mathcal{F}$ with $\lim_q \mu(B_q^2) = 1$, a subsequence (g_n^2) of (g_n^1), $f_\infty^2 \in \mathcal{L}_E^1(\Omega, \mathcal{F}, \mu)$ and $\varphi_\infty^2 \in \mathcal{L}_{\mathbf{R}}^1(\Omega, \mathcal{F}, \mu)$ such that the following hold

$$\lim_{n\to\infty} ||\frac{1}{n}\Sigma_{i=1}^n h_i(\omega) - f_\infty^2(\omega)|| = 0, \quad \text{a.e. } \omega \in A_{p_2}$$

for every subsequence (h_n) of (g_n^2) and

$$\forall q, \forall A \in \mathcal{F}, \quad \lim_{n\to\infty} \int_{A\cap A_{p_2}\cap B_q^2} ||g_n^2|| d\mu = \int_{A\cap A_{p_2}\cap B_q^2} \varphi_\infty^2 d\mu.$$

Repeating the preceding arguments provides a measurable set A_{p_k} with $\mu(A_{p_k}) > 1 - \frac{1}{k+1}$, an increasing sequence (B_q^k) in $\mathcal{F}$ with $\lim_q \mu(B_q^k) = 1$, a subsequence (g_n^k) of (g_n^{k-1}), $f_\infty^k \in \mathcal{L}_E^1(\Omega, \mathcal{F}, \mu)$ and $\varphi_\infty^k \in \mathcal{L}_{\mathbf{R}}^1(\Omega, \mathcal{F}, \mu)$ such that the following hold

$$\lim_n ||\frac{1}{n}\Sigma_{i=1}^n h_i(\omega) - f_\infty^k(\omega)|| = 0, \quad \text{a.e. } \omega \in A_{p_k}$$

for every subsequence (h_n) of (g_n^k) and

$$\forall q, \forall A \in \mathcal{F}, \quad \lim_{n\to\infty} \int_{A\cap A_{p_k}\cap B_q^k} ||g_n^k|| d\mu = \int_{A\cap A_{p_k}\cap B_q^k} \varphi_\infty^k d\mu.$$

Finally, define

$$A'_{p_1} := A_{p_1} \quad \text{and} \quad A'_{p_k} := A_{p_k} \setminus A_{p_{k-1}} \text{ for } k > 1$$

$$g_n := g_n^n, \quad f_\infty := \sum_{k=1}^{k=\infty} 1_{A'_{p_k}} f_\infty^k \quad \text{and} \quad \varphi_\infty := \sum_{k=1}^{k=\infty} 1_{A'_{p_k}} \varphi_\infty^k.$$

Then

$$\lim_{k\to\infty} \mu(A'_{p_k}) = \mu\left(\bigcup_{k=1}^{k=\infty} A'_{p_k}\right) = \mu\left(\bigcup_{k=1}^{k=\infty} A_{p_k}\right) = \lim_{k\to\infty} \mu(A_{p_k}) = 1$$

because $\mu(A_{p_k}) > 1 - \frac{1}{k+1}$ for all $k \in \mathbf{N}$. Further, it is not difficult to see that

$$\lim_{n\to\infty} ||\frac{1}{n}\Sigma_{i=1}^{n} h_i(\omega) - f_\infty(\omega)|| = 0, \quad \text{a.e. } \omega \in \Omega \tag{2.1.1}$$

for every subsequence (h_n) of (g_n) and

$$\forall q, \forall k, \forall A \in \mathcal{F}, \quad \lim_{n\to\infty} \int_{A\cap A_{p_k}\cap B_q^k} ||g_n|| d\mu = \int_{A\cap A_{p_k}\cap B_q^k} \varphi_\infty d\mu. \tag{2.1.2}$$

Now, let us prove the second part of Theorem 2.1. Applying the new condition to the sequence (g_n) provides a sequence (r_n) of the form $r_n = \sum_{i\in I_n} \lambda_i^n ||g_{i+n}||$ with $\lambda_i^n \geq 0$ and $\sum_{i\in I_n} \lambda_i^n = 1$ such that $\limsup_n r_n \in \mathcal{L}_{\mathbf{R}}^1(\Omega, \mathcal{F}, \mu)$. From (2.1.2) and Fatou lemma it follows that

$$\begin{aligned}
\int_{A'_{p_k}\cap B_q^k} \varphi_\infty d\mu &= \lim_{n\to\infty} \sum_{i\in I_n} \lambda_i^n \int_{A'_{p_k}\cap B_q^k} ||g_{i+n}|| d\mu \\
&= \lim_{n\to\infty} \int_{A'_{p_k}\cap B_q^k} \sum_{i\in I_n} \lambda_i^n ||g_{i+n}|| d\mu \\
&\leq \int_{A'_{p_k}\cap B_q^k} \limsup_{n\to\infty} r_n d\mu.
\end{aligned} \tag{2.1.3}$$

Since the sequence $(\frac{1}{n}\Sigma_{i=1}^n g_{n+i})$ is uniformly integrable on each $A_{p_k} \cap B_q^k$, it follows from (2.1.1) and Lebesgue–Vitali theorem that

$$\lim_{n\to\infty} \int_{A'_{p_k}\cap B_q^k} ||\frac{1}{n}\Sigma_{i=1}^n g_{n+i}|| d\mu = \int_{A'_{p_k}\cap B_q^k} ||f_\infty|| d\mu. \tag{2.1.4}$$

Therefore, by (2.1.4), (2.1.2) and (2.1.3) we deduce that

$$\begin{aligned}
\int_{A'_{p_k}\cap B_q^k} ||f_\infty|| d\mu &\leq \lim_n \frac{1}{n}\Sigma_{i=1}^n \int_{A'_{p_k}\cap B_q^k} ||g_{n+i}|| d\mu \\
&= \lim_n \int_{A'_{p_k}\cap B_q^k} ||g_n|| d\mu \\
&= \int_{A'_{p_k}\cap B_q^k} \varphi_\infty d\mu \\
&\leq \int_{A'_{p_k}\cap B_q^k} \limsup_{n\to\infty} r_n d\mu.
\end{aligned}$$

Then it follows that

$$\begin{aligned}\int_\Omega ||f_\infty||d\mu &= \Sigma_{k=1}^{+\infty} \lim_{q\to\infty} \int_{A'_{p_k}\cap B^k_q} ||f_\infty||d\mu \\ &\le \Sigma_{k=1}^{+\infty} \lim_{q\to\infty} \int_{A'_{p_k}\cap B^k_q} \limsup_{n\to\infty} r_n d\mu \\ &= \int_\Omega \limsup_{n\to\infty} r_n d\mu < +\infty.\end{aligned}$$

The proof is therefore complete. □

Before going further let us mention a useful convergence result.

Proposition 2.1. *Let $\mathcal{H}$ be a subset of $\mathcal{L}^l_{\mathbf{R}}(\Omega, \mathcal{F}, \mu)$ with $\ell = 0, 1$. Then the following are equivalent:*
(1) Given any sequence (f_n) in $\mathcal{H}$, there exist a subsequence (g_n) of (f_n), φ_∞ in $\mathcal{L}^\ell_{\mathbf{R}^+}(\Omega, \mathcal{F}, \mu)$ and an increasing sequence (C_k) in $\mathcal{F}$ with $\lim_{k\to\infty}\mu(C_k) = 1$ *such that for every $k \in \mathbf{N}$, $1_{C_k}\varphi_\infty \in \mathcal{L}^1_{\mathbf{R}^+}(\Omega, \mathcal{F}, \mu)$ and*

$$\forall A \in \mathcal{F}, \quad \lim_{n\to\infty} \int_{A\cap C_k} |g_n| d\mu = \int_{A\cap C_k} \varphi_\infty d\mu.$$

(2) Given any sequence (f_n) in $\mathcal{H}$ there exists a sequence (r_n) in $\mathcal{L}^1_{\mathbf{R}}(\Omega, \mathcal{F}, \mu)$ with $r_n \in co\{|f_i| : i \ge n\}$ such that (r_n) pointwisely converges a.e to a measurable function $r_\infty \in \mathcal{L}^\ell_{\mathbf{R}}(\Omega, \mathcal{F}, \mu)$.
(3) Given any sequence (f_n) in $\mathcal{H}$, there exists a sequence (r_n) in $\mathcal{L}^0_{\mathbf{R}}(\Omega, \mathcal{F}, \mu)$ with $r_n \in co\{|f_i| : i \ge n\}$ such that $\limsup_n r_n \in \mathcal{L}^\ell_{\mathbf{R}}(\Omega, \mathcal{F}, \mu)$.

Proof. Suppose (1) holds. Let (f_n) be a sequence in $\mathcal{H}$. There exist a subsequence (g_n) of (f_n), $\varphi_\infty \in \mathcal{L}^\ell_{\mathbf{R}^+}(\Omega, \mathcal{F}, \mu)$ and an increasing sequence (C_k) in $\mathcal{F}$ with $\lim_{k\to\infty}\mu(C_k) = 1$ such that

$$\forall k, \forall A \in \mathcal{F}, \quad \lim_{n\to\infty} \int_{A\cap C_k} |g_n| d\mu = \int_{A\cap C_k} \varphi_\infty d\mu < \infty.$$

For each k choose n_k such that, for every $n \ge n_k$, $\int_{C_k} |g_n| d\mu < +\infty$ and define the following sequence of integrable functions:

$$\begin{aligned} g_{n,k} &= 1_{C_k} g_n \quad \text{if } n \ge n_k, \\ &= 0 \quad \text{otherwise.}\end{aligned}$$

It is clear that, for each k, the sequence $(|g_{n,k}|)_n$ $\sigma(L^1, L^\infty)$-converges to $1_{C_k}\varphi_\infty$. By Lemma 3.1 in [17], there exists a sequence $(r_{n,k})_{n,k}$ with $r_{n,k} =$

$\sum_{i=n}^{\nu_n} \lambda_i^n |g_{i,k}|$ where $\lambda_i^n \geq 0$ and $\sum_{i=n}^{\nu_n} = 1$ such that, for every k, $(r_{n,k})_n$ converges a.e. to $1_{C_k}\varphi_\infty$. Since

$$\forall k, \forall \omega \in C_k, \quad \lim_{n\to\infty} \sum_{i=n}^{\nu_n} \lambda_i^n |g_{i,k}| = \lim_{n \geq n_k, n\to\infty} \sum_{i=n}^{\nu_n} \lambda_i^n |g_i| = \lim_{n\to\infty} \sum_{i=n}^{\nu_n} \lambda_i^n |g_i|$$

and $\lim_{k\to\infty} \mu(C_k) = 1$ we deduce that the sequence $(\sum_{i=n}^{\nu_n} \lambda_i^n |g_i|)$ converges a.e to φ_∞. This proves (2). The implication (2) $\Rightarrow$ (3) is trivial. Now, to prove the implication (3) $\Rightarrow$ (1) let (f_n) be a sequence in $\mathcal{H}$ and let (g_n), φ_∞, A_{p_k} and B_q^k, $(k, q \geq 1)$, be defined as in the proof of Theorem 2.1. Since $\lim_q \mu(B_q^k) = 1$, then there exists $q_k \geq 1$ such that $\mu(B_{q_k}^k) \geq 1 - \frac{1}{k}$. Taking $C_k := \cup_{i=1}^{i=k}(A_{p_i} \cap B_{q_i}^i)$ it is clear that (g_n), φ_∞ and (C_k) have the required properties. □

There is a simple version of Theorem 2.1.

Corollary 2.1. *Let E be a separable super reflexive Banach space and let (f_n) be a sequence in $\mathcal{L}_E^0(\Omega, \mathcal{F}, \mu)$ such that* $\limsup_n ||f_n(\omega)||$ *is finite for each ω in Ω. Then there exist a subsequence (g_n) of (f_n) and an E-valued random vector f_∞ such that*

$$\lim_n ||\frac{1}{n}\sum_{i=1}^{n} h_i(\omega) - f_\infty(\omega)|| = 0, \quad a.e.\ \omega \in \Omega,$$

for every subsequence (h_n) of (g_n). If $\limsup_n ||f_n(.)|| \in \mathcal{L}_{\mathbf{R}}^1(\Omega, \mathcal{F}, \mu)$, *then $f_\infty \in \mathcal{L}_E^1(\Omega, \mathcal{F}, \mu)$.*

Proof. Corollary 2.1 is a direct consequence of Theorem 2.1, because for every subsequence (f_{n_k}) and for every sequence (r_n) in $\mathcal{L}_{\mathbf{R}}^0$ with $r_n \in co\{||f_{n_i}|| : i \geq n\}$ we have

$$\limsup_n r_n(\omega) \leq \limsup_n ||f_n(\omega)||, \ \omega \in \Omega,$$

see Lemma 4.1 in [15] or more generally Lemma 1.1 in [5]. □

Remark. Theorem 2.1 is not true for unbounded sequences, it suffices to take $f_n = n, \forall n \in \mathbf{N}$.

Now we proceed with some significant variants of Theorem 2.1 when the Banach space E is not super reflexive. We recall some *tightness* notions.

Let F be a separable Banach space. Let $cwk(F)$ (respectively $\mathcal{R}wc(F)$) be the set of all convex weakly compact (respectively, weakly closed, ball-weakly compact) subsets in F; a weakly closed subset in F is *ball-weakly compact* if its intersection with any closed ball of F is weakly compact. If F is reflexive, any weakly closed subset of F is ball-weakly compact. Similarly, any weak* closed subset of the weak* dual of a Banach space F, is ball-weakly* compact. A sequence (f_n) in $\mathcal{L}^1_F(\Omega, \mathcal{F}, \mu)$ is $\mathcal{R}wc(F)$*-tight*, if, for every $\varepsilon > 0$ there exists a $\mathcal{R}wc(F)$-valued measurable multifunction $\Gamma_\varepsilon : \Omega \rightrightarrows F$ such that

$$\sup_n \mu(\{\omega \in \Omega : f_n(\omega) \notin \Gamma_\varepsilon(\omega)\}) < \varepsilon$$

Theorem 2.2. *Let F be a separable Banach space and $D := (e'_p)_{p\geq 1}$ be a dense sequence in F' for the Mackey topology. Let (f_n) be a $\mathcal{R}wc(F)$-tight sequence in $\mathcal{L}^1_F(\Omega, \mathcal{F}, \mu)$ satisfying the condition (*): for every subsequence (f_{n_k}) of (f_n) there exists a sequence (r_n) in $\mathcal{L}^1_{\mathbf{R}}(\Omega, \mathcal{F}, \mu)$ with $r_n \in co\{||f_{n_i}(.)|| : i \geq n\}$ such that $\limsup_n r_n(\omega)$ is finite for each ω in Ω. Then there exist a subsequence (g_n) of (f_n) and an F-valued random vector f_∞ such that*

$$\forall p \geq 1, \quad \frac{1}{n}\sum_{i=1}^{n}\left\langle e'_p, h_i(\omega)\right\rangle \to \left\langle e'_p, f_\infty(\omega)\right\rangle, \quad a.e.\ \omega \in \Omega,$$

$$\forall u \in \mathcal{L}^0_{F'}[F](\Omega, \mathcal{F}, \mu), \mu\text{–}\lim_{n\to\infty} \frac{1}{n}\sum_{i=1}^{n}\langle u, h_i\rangle = \langle u, f_\infty\rangle$$

for every subsequence (h_n) of (g_n). Furthermore, if we suppose in the condition (*), $\limsup_n r_n(.) \in \mathcal{L}^1_{\mathbf{R}}(\Omega, \mathcal{F}, \mu)$, then $f_\infty \in \mathcal{L}^1_F(\Omega, \mathcal{F}, \mu)$.

Proof. Let us prove the first part of Theorem 2.2. By Proposition 2.1, there exist a subsequence (g_n) of (f_n), $\varphi_\infty \in \mathcal{L}^0_{\mathbf{R}^+}(\Omega, \mathcal{F}, \mu)$ and an increasing sequence (C_k) in $\mathcal{F}$ with $\lim_k \mu(C_k) = 1$ such that for every $k \in \mathbf{N}$

$$\lim_{n\to\infty} \int_{A\cap C_k} ||g_n|| d\mu = \int_{A\cap C_k} \varphi_\infty d\mu. \tag{2.2.1}$$

for all $A \in \mathcal{F}$. It is clear that, for each $k \geq 1$, the sequence $(1_{C_k} g_n)$ is uniformly integrable and $\mathcal{R}wc(F)$-tight in $\mathcal{L}^1_F(\Omega, \mathcal{F}, \mu)$. Using (2.2.1) and Corollary 2.1 in [23] via an appropriate diagonal procedure, we find a subsequence of (g_n) still denoted (g_n) and $f^k_\infty \in \mathcal{L}^1_F(\Omega, \mathcal{F}, \mu)$ such that

$$\forall k \geq 1, \forall p \geq 1, \quad \lim_{n\to\infty} \frac{1}{n}\sum_{i=1}^{n}\langle e'_p, 1_{C_k} h_i(\omega)\rangle = \langle e'_p, f^k_\infty(\omega)\rangle, \quad \text{a.e. } \omega \in \Omega$$

$$\forall k \geq 1, \forall u \in \mathcal{L}^0_{F'}[F](\Omega, \mathcal{F}, \mu), \quad \mu - \lim_{n\to\infty} \frac{1}{n}\sum_{i=1}^{n}\langle u, 1_{C_k} h_i\rangle = \langle u, f^k_\infty\rangle$$

for every subsequence (h_n) of (g_n). Putting

$$C'_1 := C_1 \quad \text{and} \quad C'_k := C_k \setminus \bigcup_{i<k} C_i \ \text{ for } k > 1$$

and

$$f_\infty := \sum_{k=1}^{k=\infty} 1_{C'_k} f_\infty^k,$$

it is not difficult to verify that

$$\forall p \geq 1, \quad \lim_{n\to\infty} \frac{1}{n} \sum_{i=1}^{n} \left\langle e'_p, h_i(\omega) \right\rangle = \left\langle e'_p, f_\infty(\omega) \right\rangle, \quad \text{a.e. } \omega \in \Omega \; (2.2.2)$$

$$\forall u \in \mathcal{L}^0_{F'}[F](\Omega, \mathcal{F}, \mu), \quad \mu\text{-} \lim_{n\to\infty} \frac{1}{n} \sum_{i=1}^{n} \langle u, h_i \rangle = \langle u, f_\infty \rangle \qquad (2.2.3)$$

for every subsequence (h_n) of (g_n).

Now, let us prove the second part of Theorem 2.2. Applying the new condition to the sequence (g_n) provides a sequence (r_n) of the form $r_n = \sum_{i\in I_n} \lambda_i^n ||g_{i+n}||$ with $\lambda_i^n \geq 0$ and $\sum_{i\in I_n} \lambda_i^n = 1$ such that $\limsup_n r_n(.) \in \mathcal{L}^1_{\mathbf{R}}(\Omega, \mathcal{F}, \mu)$. From (2.2.1) and Fatou lemma it follows that

$$\begin{aligned}
\int_{C'_k} \varphi_\infty d\mu &= \lim_n \sum_{i\in I_n} \lambda_i^n \int_{C'_k} ||g_{i+n}|| d\mu \\
&= \lim_n \int_{C'_k} \sum_{i\in I_n} \lambda_i^n ||g_{i+n}|| d\mu \\
&\leq \int_{C'_k} \limsup_n r_n d\mu. \qquad (2.2.4)
\end{aligned}$$

On the other hand, since the sequence $(1_{C_k} g_n)$ is uniformly integrable, from (2.2.3) and Lebesgue–Vitali it follows that (2.2.5)

$$\forall k \geq 1, \forall u \in L^\infty_{F'}[F](\Omega, \mathcal{F}, \mu), \quad \lim_{n\to\infty} \int_{C'_k} \langle u, \frac{1}{n} \Sigma_{i=1}^n g_{i+k} \rangle d\mu = \int_{C'_k} \langle u, f_\infty \rangle d\mu.$$

Since the norm $||.||_{L^1_F}$ is weakly lower semi-continuous, (2.2.5) implies

$$\begin{aligned}
\forall k \geq 1, \quad \int_{C'_k} ||f_\infty|| d\mu &\leq \liminf_{n\to\infty} \int_{C'_k} ||\frac{1}{n} \Sigma_{i=1}^n g_{i+k}|| d\mu \\
&\leq \liminf_{n\to\infty} \int_{C'_k} \frac{1}{n} \Sigma_{i=1}^n ||g_{i+k}|| d\mu \\
&= \lim_{n\to\infty} \int_{C'_k} ||g_n|| d\mu \\
&= \int_{C'_k} \varphi_\infty d\mu
\end{aligned}$$

where the two last inequalities follow from (2.2.1). Thus the above estimate and (2.2.4) imply that

$$\forall k \geq 1, \quad \int_{C'_k} ||f_\infty|| d\mu \leq \int_{C'_k} \limsup_n r_n d\mu.$$

Hence

$$\begin{aligned}\int_\Omega ||f_\infty|| d\mu &= \Sigma_{k=1}^{+\infty} \int_{C'_k} ||f_\infty|| d\mu \\ &\leq \Sigma_{k=1}^{+\infty} \int_{C'_k} \limsup_n r_n d\mu \\ &= \int_\Omega \limsup_n r_n d\mu < +\infty.\end{aligned}$$

□

As a special case of Theorem 2.2 we give the following result.

Corollary 2.2. *Let F be a separable Banach space and let (f_n) be a sequence in $\mathcal{L}^1_F(\Omega, \mathcal{F}, \mu)$ satisfying the condition (*) of Theorem 2.2 and the condition (**): there exists a measurable $\mathcal{R}wc(F)$-valued multifunction $\Gamma : \Omega \rightrightarrows F$ such that $f_n(\omega) \in \Gamma(\omega)$ for all $n \in \mathbf{N}$ and all $\omega \in \Omega$. Then there exist a subsequence (g_n) of (f_n) and an E-valued random vector f_∞ such that*

$$\forall x^* \in F', \quad \lim_{n\to\infty} \frac{1}{n} \sum_{i=1}^{n} \langle x^*, h_i(\omega)\rangle \to \langle x^*, f_\infty(\omega)\rangle, \quad a.e.\ \omega \in \Omega$$

for every subsequence (h_n) of (g_n), here the negligible set depend only the subsequence under consideration. Furthermore, if we suppose in the condition (*) $\limsup_n r_n(.) \in \mathcal{L}^1_{\mathbf{R}}(\Omega, \mathcal{F}, \mu)$*, then $f_\infty \in \mathcal{L}^1_E(\Omega, \mathcal{F}, \mu)$.*

Proof. It clear that (f_n) is $\mathcal{R}wc(F)$-tight. Applying Theorems 2.1 and 2.2, respectively, to the sequences $(||f_n||)_n$ and $(\langle e'_k, f_n\rangle)_n$ $(k \geq 1)$ via a diagonal process provides a subsequence (g_n) of (f_n) and $f_\infty \in \mathcal{L}^0_F(\Omega, \mathcal{F}, \mu)$, (or $\in \mathcal{L}^1_F(\Omega, \mathcal{F}, \mu)$, if we suppose in condition (*), $\limsup_n r_n(.) \in \mathcal{L}^1_F(\Omega, \mathcal{F}, \mu)$) such that

$$\lim_{n\to\infty} \frac{1}{n} \Sigma_{i=1}^n ||h_i(\omega)|| \text{ exists}, \quad \text{a.e. } \omega \in \Omega, \tag{2.2.5}$$

$$\forall k \geq 1, \quad \lim_{n\to\infty} \frac{1}{n} \sum_{i=1}^{n} \langle e'_k, h_i(\omega)\rangle = \langle e'_k, f_\infty(\omega)\rangle, \quad \text{a.e. } \omega \in \Omega, \tag{2.2.6}$$

for every subsequence (h_n) of (g_n). To complete the proof take a subsequence (h_n) of (f_n) and put $\Gamma(.) := R(.) \cap r(.)\overline{B}_E$, where $r(.) := \sup_n \frac{1}{n} \Sigma_{i=1}^n ||h_i(.)||$.

Then, by (2.2.5), $\Gamma(\omega)$ is weakly compact a.e and $\forall n \in \mathbf{N}$, $\frac{1}{n}\Sigma_{i=1}^{n} h_i(\omega) \in \overline{co}\Gamma(\omega)$ a.e. Hence, it follows from (2.2.6), by using a routine density argument, that

$$\forall x^* \in F', \quad \lim_{n\to\infty} \frac{1}{n}\sum_{i=1}^{n}\langle x^*, h_i(\omega)\rangle = \langle x^*, f_\infty(\omega)\rangle, \quad \text{a.e. } \omega \in \Omega.$$

□

Remark. Theorem 2.2 is valid in the space $\mathcal{L}_F^0(\Omega, \mathcal{F}, \mu)$ under obvious modifications.

Now we present some convergence properties for a class of unbounded sequences in the space $L_{F'}^1[F](\Omega, \mathcal{F}, \mu)$ of all F'-valued mappings $f : \Omega \to F'$ such that $\omega \mapsto \langle f(\omega), x\rangle$ is integrable, $\forall x \in F$, and such that $|f|(.) := ||f(.)||_{F'}$ belongs to $L_{\mathbf{R}}^1(\Omega, \mathcal{F}, \mu)$, when F is a separable Banach space. We summarize some properties of this space. For more details, we refer to [5,18] and the references therein. We will endow $L_{F'}^1[F](\Omega, \mathcal{F}, \mu)$ with the norm

$$\overline{N}_1(f) = N_1(||f||), \quad f \in L_{F'}^1[F](\Omega, \mathcal{F}, \mu)$$

here N_1 denotes the norm in $L_{\mathbf{R}}^1(\Omega, \mathcal{F}, \mu)$. By Theorem 4.1 in [5] $L_F^\infty(\Omega, \mathcal{F}, \mu)$ is included in the topological dual $(L_{F'}^1[F](\Omega, \mathcal{F}, \mu))'$ of $L_{F'}^1[F](\mu)$ and we have

$$\overline{N}_1(f) = \sup_{h\in\mathcal{H}_1} \int_\Omega \langle h, f\rangle d\mu$$

here $\mathcal{H}_1$ denotes the set of all simple mappings from Ω into the closed unit ball $\overline{B}_F$ of F, so that the mapping $f \mapsto \overline{N}_1(f)$ is lower semicontinuous on $L_{F'}^1[F](\Omega, \mathcal{F}, \mu)$ for the topology $\sigma(L_{F'}^1[F](\mu), L_F^\infty(\mu))$.

The following result deals with some convergence properties for a class of unbounded sequences in $(L_{F'}^1[F](\Omega, \mathcal{F}, \mu), \overline{N}_1)$ and leads to interesting applications in several problems of convergence of F'-valued scalarly integrable random variables, in particular, Fatou type Lemma in $L_{F'}^1[F](\Omega, \mathcal{F}, \mu)$.

Theorem 2.3. *Let F be a separable Banach space. Let (f_n) be a sequence in $(L_{F'}^1[F](\Omega, \mathcal{F}, \mu), \overline{N}_1)$ which satisfies the condition (*): for every subsequence (f_{n_k}) of (f_n) there exists a sequence (r_n) in $L_{\mathbf{R}}^1(\Omega, \mathcal{F}, \mu)$ with $r_n \in co\{|f_{n_i}| : i \geq n\}$ such that* $\limsup_n r_n \in L_{\mathbf{R}}^1(\Omega, \mathcal{F}, \mu)$. *Then there exist a subsequence (g_n) of (f_n), $f_\infty \in L_{F'}^1[F](\Omega, \mathcal{F}, \mu)$ and an increasing sequence (C_k) in $\mathcal{F}$ with $\lim_k \mu(C_k) = 1$ such that*

$$\forall k \geq 1, \forall v \in L_F^\infty(\Omega, \mathcal{F}, \mu), \quad \lim_{n\to\infty} \int_{C_k} \langle v, g_n\rangle d\mu = \int_{C_k} \langle v, f_\infty\rangle d\mu$$

and such that

$$f_\infty(\omega) \in \overline{co}\left(\bigcap_n w^*\text{-}cl\{f_m(\omega) : m \geq n\}\right), \quad a.e.$$

Furthermore, we have

$$\forall x \in F, \quad \lim_{n\to\infty} \frac{1}{n}\sum_{i=1}^{n}\langle x, h_i\rangle = \langle x, f_\infty\rangle, \ a.e.$$

for every subsequence (h_n) of (g_n), here the negligible set depends only on the subsequence under consideration.

Proof. Step 1 Applying Proposition 2.1 to the sequence $(|f_n|)$ provides a subsequence (g_n) of (f_n), $\varphi_\infty \in L^1_{\mathbf{R}^+}(\Omega, \mathcal{F}, \mu)$ and an increasing sequence (C_k) in $\mathcal{F}$ with $\lim_k \mu(C_k) = 1$ such that for every $k \in N$

$$\lim_{n\to\infty} \int_{A\cap C_k} ||g_n|| d\mu = \int_{A\cap C_k} \varphi_\infty d\mu. \tag{2.3.1}$$

for all $A \in \mathcal{F}$. Then for every $k \in \mathbf{N}$, $(1_{C_k} g_n)_{n\in\mathbf{N}}$ is uniformly integrable in $L^1_{F'}[F](C_k, C_k \cap \mathcal{F}, \mu|_{C_k})$. In view of Theorem 6.5.9 in [13] and a diagonal procedure, we produce a subsequence (not relabelled) of (g_n) and a sequence (f^k_∞) with $f^k_\infty \in L^1_{F'}[F](C_k, C_k \cap \mathcal{F}, \mu|_{C_k})$ such that

$$f^k_\infty(\omega) \in \overline{co}\left(\bigcap_p w^*\text{-}cl\{f_m(\omega) : m \geq p\}\right), \quad \text{a.e.} \quad \omega \in C_k \tag{2.3.2}$$

and such that

$$\forall k \in \mathbf{N}, \forall v \in L^\infty_F(\Omega, \mathcal{F}, \mu), \quad \lim_{n\to\infty} \int_{C_k} \langle v, g_n\rangle d\mu = \int_{C_k} \langle v, f^k_\infty\rangle d\mu.$$

Define

$$C'_1 := C_1 \quad \text{and} \quad C'_k := C_k \setminus \bigcup_{i<k} C_i \ \text{ for } k > 1$$

and

$$f_\infty := \sum_{k=1}^{k=\infty} 1_{C'_k} f^k_\infty.$$

It follows that

$$\forall k \in \mathbf{N}, \forall u \in L^{\infty}_{F'}[F](\Omega, \mathcal{F}, \mu), \quad \lim_{n\to\infty} \int_{C'_k} \langle u, g_n\rangle d\mu = \int_{C'_k} \langle u, f_\infty\rangle d\mu. \tag{2.3.3}$$

On the other hand, let $D := (e_p)_{p\geq 1}$ be a dense sequence in F for the norm topology. Applying Theorem 2.1, respectively, to the sequences $(\langle e_p, g_n(.)\rangle)_n$, $(p \in \mathbf{N})$, and $(||g_n(.)||)_n$ via a standard diagonal extraction procedure provides a subsequence (not relabelled) of (g_n) such that

$$\forall p \in \mathbf{N}, \lim_{n\to\infty} \frac{1}{n}\sum_{i=1}^{n} \langle e_p, h_i(\omega)\rangle \quad \text{exists a.e.} \tag{2.3.4}$$

and

$$\lim_{n\to\infty} \frac{1}{n}\sum_{i=1}^{n} ||h_i(\omega)|| \quad \text{exists a.e.} \tag{2.3.5}$$

for every subsequence (h_n) of (g_n). By (2.3.5) the sequence $(\frac{1}{n}\sum_{i=1}^{n} ||h_n||)$ is pointwise bounded almost everywhere. By (2.3.3) and (2.3.4) it is immediate that

$$\forall p \in \mathbf{N}, \quad \lim_{n\to\infty} \frac{1}{n}\sum_{i=1}^{n} \langle e_p, h_i(\omega)\rangle = \langle e_p, f_\infty\rangle.$$

Using the separability of F and the pointwise boundedness of $(\frac{1}{n}\sum_{i=1}^{n} ||h_n||)$ we get, by a routine argument

$$\lim_{n\to\infty} \frac{1}{n}\sum_{i=1}^{n} \langle e, h_i\rangle \to \langle e, f_\infty\rangle$$

for all $e \in F$ and almost everywhere.
Step 2 $f_\infty \in L^1_{F'}[F](\Omega, \mathcal{F}, \mu)$. As

$$\overline{N}_1(f) = \sup_{h\in\mathcal{H}_1} \int_\Omega \langle h, f\rangle d\mu = \int_\Omega ||f(.)||d\mu, \quad \forall f \in L^1_{E'}[E](\mu)$$

$\overline{N}_1$ is $\sigma(L^1_{F'}[F], L^\infty_F)$ lower semicontinuous, (2.3.3) implies

$$\forall k \geq 1, \quad \int_{C'_k} ||f_\infty||d\mu \leq \liminf_{n\to\infty} \int_{C'_k} ||g_n(.)||d\mu$$
$$= \lim_{n\to\infty} \int_{C'_k} ||g_n||d\mu = \int_{C'_k} \varphi_\infty d\mu$$

where the two last inequalities follow from (2.3.1). Hence

$$\int_\Omega ||f_\infty|| d\mu = \Sigma_{k=1}^{+\infty} \int_{C'_k} ||f_\infty|| d\mu$$
$$\leq \Sigma_{k=1}^{+\infty} \int_{C'_k} \varphi_\infty d\mu = \int_\Omega \varphi_\infty d\mu < +\infty.$$

Finally by (2.3.2) we have

$$f_\infty(\omega) \in \overline{co}\left(\bigcap_n w^*\text{-}cl\{f_m(\omega) : m \geq n\}\right), \quad \text{a.e.}$$

□

Remarks. If $(|f_n|)$ is bounded, condition (*) is satified. See for instance Proposition 2.1.

Now we proceed to a new version of Komlós slice theorem in the space $\mathcal{L}^0_{cwk(E)}(\Omega, \mathcal{F}, \mu)$. Compare with ([17], Theorem 4.1 and Corollary 4.2). See [4, 12] for other related results. Let us recall the following result.

Proposition 2.2. Hess [9] *Let F be a separable Banach space, and $D = (e'_k)_{k\geq 1}$ be a dense sequence in F' for the Mackey topology. Let (X_n) be a sequence in $\mathcal{L}^0_{cwk(F)}(\Omega, \mathcal{F})$. Assume that the two following conditions are satisfied:*

(i) There is a convex weakly compact valued random set C such that, $\forall n \geq 1, \forall \omega \in \Omega, X_n(\omega) \subset C(\omega)$.
(ii) $\forall k \geq 1, \lim_{n\to\infty} \delta^*(e'_k, X_n(\omega))$ exists a.e.

Then there exists a convex weakly compact valued random set X_∞ and a negligible set $N \subset \Omega$ such that

$$\lim_{n\to\infty} \delta^*(e', X_n(\omega)) = \delta^*(e', X_\infty(\omega))$$

for all $e' \in F'$ and for all $\omega \in \Omega \setminus N$.

Theorem 2.4. *Let E be a separable super reflexive Banach space. Let (X_n) be a sequence of convex weakly compact random sets in E such that for every subsequence (X_{n_k}) of (X_n) there exists a sequence (r_n) in $\mathcal{L}^0_{\mathbf{R}}(\Omega, \mathcal{F}, \mu)$ with $r_n \in co\{|X_{n_i}| : i \geq n\}$ such that $\limsup_n r_n(\omega)$ is finite for each ω in Ω. Then there exist a subsequence $(X_{\beta(n)})$ of (X_n) and a convex weakly compact random set X_∞ such that, for all bounded closed convex subset B in E and for every subsequence $(X_{\gamma(n)})$ of $(X_{\beta(n)})$, the following hold:*

$$\lim_{n\to\infty} D\left(B, \frac{1}{n}\Sigma_{j=1}^n X_{\gamma(j)}(\omega)\right) = D(B, X_\infty(\omega)) \qquad (*)$$

for almost every $\omega \in \Omega$.

Proof. Let $D'_1 := (e^*_k)_{k\geq 1}$ be a dense sequence in $\overline{B}_{E^*}$ for the norm topology. For each k, we pick a maximum $\mathcal{F}$-measurable selection σ^k_n of X_n associated to e^*_k, that is, $\langle e^*_k, \sigma^k_n\rangle = \delta^*(e^*_k, X_n)$. By virtue of Theorem 2.1, there are a subsequence $(X_{\beta(n)})$, a subsequence $(\sigma^k_{\beta(n)})$ and $\sigma^k_\infty \in \mathcal{L}^0_E(\Omega, \mathcal{F}, \mu)$ and a real valued $\mathcal{F}$-measurable function φ^k_∞ such that for each k the following hold

$$\lim_{n\to\infty} \delta^*\left(e^*_k, \frac{1}{n}\Sigma^n_{j=1}X_{\gamma(j)}(\omega)\right) = \varphi^k_\infty(\omega) \text{ a.e.} \tag{2.4.1}$$

for every subsequence $(X_{\gamma(n)})$ of $(X_{\beta(n)})$, and

$$\lim_{n\to\infty} \frac{1}{n}\Sigma^n_{j=1}\sigma^k_{\gamma(j)}(\omega) = \sigma^k_\infty(\omega) \text{ a.e.} \tag{2.4.2}$$

for every subsequence $(\sigma^k_{\gamma(n)})$ of $(\sigma^k_{\beta(n)})$. For simplicity we set

$$S_n(\omega) := \frac{1}{n}\Sigma^n_{j=1}X_{\gamma(j)}(\omega)$$

for all $n \in \mathbf{N}$ and for all $\omega \in \Omega$. Again by Theorem 2.1, we may assume that for each $\omega \in \Omega$,

$$\sup_n |S_n(\omega)| < +\infty \tag{2.4.3}$$

a.e. By (2.4.1), (2.4.3) and Proposition 2.2, there is a convex weakly random set X_∞ and a negligible set N such that

$$\lim_{n\to\infty} \delta^*\left(e^*, \frac{1}{n}\Sigma^n_{j=1}X_{\gamma(j)}(\omega)\right) = \delta^*\left(e^*, X_\infty(\omega)\right) \tag{2.4.4}$$

for all $(\omega, e^*) \in (\Omega \setminus N) \times \overline{B}_{E^*}$. Now by obvious properties of (σ^k_n), (2.4.1), (2.4.2) and the definition of $s\text{-}li\frac{1}{n}\Sigma^n_{j=1}X_{\gamma(j)}(\omega)$ we have

$$\delta^*\left(e^*_k, X_\infty(\omega)\right) = \left\langle e^*_k, \sigma^k_\infty(\omega)\right\rangle \leq \delta^*\left(e^*_k, s\text{-}li\frac{1}{n}\sum_{j=1}^n X_{\gamma(j)}(\omega)\right) \tag{2.4.5}$$

for all k and almost every $\omega \in \Omega$. By (2.4.4), (2.4.5) we conclude that (S_n) Mosco-converges to X_∞. We will prove (*) that is a formulation of slice convergence in $\mathcal{L}^0_{cwk(E)}(\Omega, \mathcal{F}, \mu)$. By (2.4.4) we have

$$\begin{aligned}
\liminf_{n\to\infty} D(B, S_n(\omega)) &= \liminf_{n\to\infty} \sup_{x^*\in\overline{B}_{E'}} \left\{-\delta^*(x^*, S_n(\omega)) - \delta^*(-x^*, B)\right\} \\
&\geq \sup_{x^*\in\overline{B}_{E'}} \liminf_{n\to\infty} \left\{-\delta^*(x^*, S_n(\omega)) - \delta^*(-x^*, B)\right\} \\
&= \sup_{x^*\in\overline{B}_{E'}} \left\{- \limsup_{n\to\infty} \delta^*(x^*, S_n(\omega)) - \delta^*(-x^*, B)\right\} \\
&= \sup_{x^*\in\overline{B}_{E'}} \left\{-\delta^*(x^*, X_\infty(\omega)) - \delta^*(-x^*, B)\right\} \\
&= D(B, X_\infty(\omega))
\end{aligned}$$

for any bounded closed convex subset B in E and for almost every $\omega \in \Omega$. This proves the *liminf part* .

Let us prove now the *limsup part* . Let (g_k) be a Castaing's representation of $s\text{-}li\frac{1}{n}\Sigma_{j=1}^n X_{\gamma(j)}(.)$. Then we have

$$\forall\omega \in \Omega, \quad \forall x \in E, \quad \forall k \in \mathbf{N}, \quad \limsup_{n\to\infty} d\left(x, \frac{1}{n}\Sigma_{j=1}^n X_{\gamma(j)}(\omega)\right) \leq d(x, g_k(\omega))$$

which implies that

$$\begin{aligned}
\limsup_{n\to\infty} d\left(x, \frac{1}{n}\Sigma_{j=1}^n X_{\gamma(j)}(\omega)\right) &\leq \inf_{k\in\mathbf{N}} d(x, g_k(\omega)) \\
&= d\left(x, s\text{-}li\frac{1}{n}\Sigma_{j=1}^n X_{\gamma(j)}(\omega)\right).
\end{aligned}$$

Now let B be a bounded closed convex subset of E, then

$$\begin{aligned}
\limsup_{n\to\infty} D\left(B, \frac{1}{n}\Sigma_{j=1}^n X_{\gamma(j)}(\omega)\right) &= \limsup_{n\to\infty} \inf_{x\in B} d(x, \frac{1}{n}\Sigma_{j=1}^n X_{\gamma(j)}(\omega)) \\
&\leq \inf_{x\in B} \limsup_{n\to\infty} d\left(x, \frac{1}{n}\Sigma_{j=1}^n X_{\gamma(j)}(\omega)\right) \\
&\leq \inf_{x\in B} d\left(x, s\text{-}li\frac{1}{n}\Sigma_{j=1}^n X_{\gamma(j)}(\omega)\right) \\
&= D\left(B, s\text{-}li\frac{1}{n}\Sigma_{j=1}^n X_{\gamma(j)}(\omega)\right) \\
&\leq D(B, X_\infty(\omega))
\end{aligned}$$

for any bounded closed convex subset B in E and for almost every $\omega \in \Omega$. □

We give some applications of the preceding results.

Proposition 2.3. *Let E be a separable super reflexive Banach space. Let $\mathcal{H}$ be a convex subset of $\mathcal{L}_E^0(\Omega, \mathcal{F}, \mu)$. Then the following are equivalent:*

(a) $\mathcal{H}$ *is closed for the topology of convergence in measure,*
(b) For any almost everywhere pointwise bounded sequence (u_n) *in* $\mathcal{H}$*, that is,*

$$\sup_n ||u_n(\omega)|| < +\infty$$

for almost every $\omega \in \Omega$*, there is a subsequence* $(u_{\gamma(n)})$ *and* $u_\infty \in \mathcal{L}^0_E(\Omega, \mathcal{F}, \mu)$ *such that*

$$\mu - \lim_{n\to\infty} \frac{1}{n} \Sigma^n_{j=1} u_{\gamma(j)} = u_\infty$$

with respect to the norm topology.

Proof. $(a) \Longrightarrow (b)$. In view of Theorem 2.1, any almost everywhere pointwise bounded sequence (u_n) in $\mathcal{H}$ has the Komlós property. Hence there exist a subsequence $(u_{\gamma(n)})$ and $u_\infty \in \mathcal{L}^0_E(\Omega, \mathcal{F}, \mu)$ such that

$$\lim_{n\to\infty} \frac{1}{n} \Sigma^n_{j=1} u_{\gamma(j)} = u_\infty$$

a.e., with respect to the norm topology. In particular, we have

$$\mu - \lim_{n\to\infty} \frac{1}{n} \Sigma^n_{j=1} u_{\gamma(j)} = u_\infty$$

with respect to the norm topology. Since $\mathcal{H}$ is convex and closed for the convergence in measure, it is immediate that $u_\infty \in \mathcal{H}$.
$(b) \Longrightarrow (a)$. Let (u_n) be a sequence in $\mathcal{H}$ which converges in measure to $\overline{u} \in \mathcal{L}^0_E(\Omega, \mathcal{F}, \mu)$. There is a subsequence of (v_n) of (u_n) which converges to $\overline{u}$ for almost every $\omega \in \Omega$. Thus the sequence (v_n) is pointwise bounded for almost every $\omega \in \Omega$. Using (b) and Theorem 2.1, there is a subsequence $(v_{\alpha(n)})$ and $v_\infty \in \mathcal{H}$ such that

$$\lim_{n\to\infty} \frac{1}{n} \Sigma^n_{j=1} v_{\alpha(j)} = v_\infty$$

a.e., with respect to the norm topology. So we have $\overline{u} = v_\infty \in \mathcal{H}$ for almost every $\omega \in \Omega$. □

The following result is a combined effort of Theorem 2.1 and Theorem 2.4. So we omit the proof.

Proposition 2.4. *Let E be a separable super reflexive Banach space. Let* (X_n) *be a sequence in* $\mathcal{L}^1_{cwk(E)}(\Omega, \mathcal{F})$ *such that*

$$\sup_n |X_n| \in \mathcal{L}^1_{\mathbf{R}}(\Omega, \mathcal{F}, \mu).$$

Then there exist a subsequence $(X_{\beta(n)})$ of (X_n) and a convex weakly compact random set X_∞ such that, for all bounded closed convex subset B in E and for every subsequence $(X_{\gamma(n)})$ of $(X_{\beta(n)})$, the following hold:

$$\lim_{n\to\infty} D\left(B, \frac{1}{n}\Sigma_{j=1}^{n} X_{\gamma(j)}(\omega)\right) = D(B, X_\infty(\omega)) \qquad (*)$$

for almost every $\omega \in \Omega$.

For the sake of transparency, let us examine first the particular case where $X_n \in L^1_E(\Omega, \mathcal{F}, \mu)$. Since E is reflexive, by Dunford–Pettis theorem, (X_n) is relatively sequentially weakly compact in $L^1_E(\Omega, \mathcal{F}, \mu)$. See [11] for further references of weak compactness in Bochner and Pettis integration. So there is a subsequence (X'_n) of (X_n) which converges weakly to an element $Y_\infty \in L^1_E(\Omega, \mathcal{F}, \mu)$ with $|Y_\infty| \le g$ and, by Mazur's lemma, there is a sequence $(\widetilde{X}_n)$ of convex combinations of (X_n) which converges strongly to Y_∞ for almost every $\omega \in \Omega$. Using Theorem 2.1 or Proposition 2.3, we get more. There is a subsequence (Y_n) of (X_n) and a random integrable E-valued vector X_∞ such that

$$\lim_n ||\frac{1}{n}\Sigma_{i=1}^{n} Z_i(\omega) - X_\infty(\omega)|| = 0, \quad \text{a.e. } \omega \in \Omega, \quad (^*)$$

for every subsequence (Z_n) of (Y_n). Let W be the unit ball of $L^\infty_{E'}(\Omega, \mathcal{F}, \mu)$. Then (*) implies

$$\lim_n \frac{1}{n}\Sigma_{i=1}^{n}\langle W(\omega), Z_i(\omega)\rangle = \langle W(\omega), X_\infty(\omega)\rangle, \quad \text{a.e. } \omega \in \Omega. \quad (^{**})$$

Integrating equality (**) gives

$$\lim_n \frac{1}{n}\Sigma_{i=1}^{n}\int_\Omega \langle W(\omega), Z_i(\omega)\rangle \mu(d\omega) = \int_\Omega \langle W(\omega), X_\infty(\omega)\rangle \mu(d\omega) \quad (^{***})$$

for every subsequence (Z_n) of (Y_n). (***) shows that (Y_n) weakly converges to X_∞ in the Banach space $L^1_E(\Omega, \mathcal{F}, \mu)$. Similarly, in the multivalued case, Proposition 2.4 provides a slice type convergence for the sequence (X_n). This result is sharper than the Mazur's τ_L convergence initiated in [17], namely, there is a sequence $(\widetilde{X}_n)$ with $\widetilde{X}_n \in co\{X_m : m \ge n\}$ (i.e. $\widetilde{X}_n$ has the form $\widetilde{X}_n = \Sigma_{i=n}^{\nu_n}\lambda_i^n X_i$ with $0 \le \lambda_i^n \le 1$, $\Sigma_{i=n}^{\nu_n}\lambda_i^n = 1$) such that $(\widetilde{X}_n)$ converges in the *linear topology* τ_L [6] to a convex weakly compact random set $X_\infty \in \mathcal{L}^1_{cwk(E)}(\Omega, \mathcal{F})$ for almost everywhere $\omega \in \Omega$, that is

$$\forall x^* \in E^*, \quad \lim_{n\to\infty} \delta^*(x^*, \widetilde{X}_n) = \delta^*(x^*, X_\infty)$$

$$\forall x \in E, \quad \lim_{n\to\infty} d(x, \widetilde{X}_n) = d(x, X_\infty)$$

for a.e. $\omega \in \Omega$. The weak convergence in the space $\mathcal{L}^1_{cwk(E)}(\Omega, \mathcal{F}, \mu)$ when E is a separable Banach space, has been extensively developed in [17], see, for instance, ([17], Theorem 3.4). At this point, let us mention that the weak convergence of (X_n) towards X_∞ in $\mathcal{L}^1_{cwk(E)}(\Omega, \mathcal{F}, \mu)$, that is,

$$\forall h \in L^\infty_{E'}(\Omega, \mathcal{F}, \mu), \quad \lim_{n\to\infty} \int_\Omega \delta^*(h, X_n)\, d\mu = \int_\Omega \delta^*(h, X_\infty)\, d\mu$$

does not follow trivially from the following one

$$\forall A \in \mathcal{F}, \forall x' \in E', \quad \lim_{n\to\infty} \int_A \delta^*(x', X_n)\, d\mu = \int_A \delta^*(x', X_\infty)\, d\mu$$

because the support function is sublinear, by constrast to the $L^1_E(\Omega, \mathcal{F}, \mu)$ case, one may use trivially a density argument, meanwhile in the space $\mathcal{L}^1_{cwk(E)}(\Omega, \mathcal{F}, \mu)$, this needs a subtile argument [17] which shows the difference between the spaces $\mathcal{L}^1_{cwk(E)}(\Omega, \mathcal{F}, \mu)$ and $L^1_E(\Omega, \mathcal{F}, \mu)$. In short the routine density argument involving $L^\infty_{E'}(\Omega, \mathcal{F}, \mu)$ cannot be applied to the case of $\mathcal{L}^1_{cwk(E)}(\Omega, \mathcal{F}, \mu)$ space. See [17], Theorem 5.3) for the weak convergence in $\mathcal{L}^1_{cwk(E)}(\Omega, \mathcal{F}, \mu)$ via Komlós type arguments. So the preceding considerations and the characterization of convex closed sets for the convergence in measure (Proposition 2.3) focus the differences when dealing with super reflexive Banach spaces.

In this vein, we present a Fatou type lemma in Mathematical Economics.

Proposition 2.5. *Let E be a separable super reflexive Banach space and let (u_n) be a sequence in $\mathcal{L}^0_E(\Omega, \mathcal{F}, \mu)$ satisfying condition (*) of Theorem 2.1. Let $\varphi : \Omega \times E \to [0, +\infty[$ be an $\mathcal{F} \otimes \mathcal{B}(E)$-measurable integrand such that, $\forall \omega \in \Omega, \varphi(\omega, .)$ is convex lower semi-continuous on E (alias convex normal integrand) and such that $I_\varphi(u) := \int_\Omega \varphi(\omega, u(\omega)) d\mu(\omega)$ is finite for every $u \in \mathcal{L}^0_E(\Omega, \mathcal{F}, \mu)$. Assume further that there exists $b \in \mathbf{R}^+$ such that $b := \lim_{n\to\infty} I_\varphi(u_n)$. Then there exist a subsequence $(u_{\gamma(n)})$ of (u_n) and $u \in \mathcal{L}^0_E(\Omega, \mathcal{F}, \mu)$ such that*

$$\lim_{n\to\infty} \frac{1}{n} \Sigma_{j=1}^n u_{\gamma(j)}(\omega) = u(\omega) \quad a.e.$$

and such that

$$b = \lim_{n\to\infty} \int_\Omega \varphi(\omega, u_n(\omega))\, \mu(d\omega) \geq \int_\Omega \varphi(\omega, u(\omega))\, \mu(d\omega).$$

Proof. In view of Theorem 2.1, there exist a subsequence $(u_{\gamma(n)})$ of (u_n) and $u \in \mathcal{L}^0_E(\Omega, \mathcal{F}, \mu)$ such that

$$\lim_{n\to\infty} \frac{1}{n} \Sigma_{j=1}^{n} u_{\gamma(j)}(\omega) = u(\omega)$$

for almost everywhere $\omega \in \Omega$. By convexity we have

$$\forall \omega, \forall n, \varphi(\omega, \frac{1}{n} \Sigma_{j=1}^{n} u_{\gamma(j)}(\omega)) \leq \Sigma_{j=1}^{n} \frac{1}{n} \varphi(\omega, u_{\gamma(j)}(\omega)).$$

Hence

$$\limsup_{n} \int_{\Omega} \varphi\left(\omega, \frac{1}{n} \Sigma_{j=1}^{n} u_{\gamma(j)}(\omega)\right) \mu(d\omega) \leq b.$$

By the lower semicontinuity of $\varphi(\omega, .)$ and by Fatou lemma, we get

$$\liminf_{n\to\infty} \int_{\Omega} \varphi\left(\omega, \frac{1}{n} \Sigma_{j=1}^{n} u_{\gamma(j)}(\omega)\right) \mu(d\omega) \geq \int_{\Omega} \varphi(\omega, u(\omega))\, \mu(d\omega)$$

which implies

$$b = \lim_{n\to\infty} \int_{\Omega} \varphi(\omega, u_n(\omega))\, \mu(d\omega) \geq \int_{\Omega} \varphi(\omega, u(\omega))\, \mu(d\omega).$$

□

We finish this section with the following closure property for the convergence in measure (compare with Proposition 5.2 in [14]).

Proposition 2.6. *Let E be a separable super reflexive Banach space, let Γ be a convex weakly compact random set and let $\mathcal{S}_\Gamma$ be the set of all $\mathcal{F}$-measurable selections of Γ. Let $\mathcal{H}$ be a convex closed in measure subset of $\mathcal{L}_E^0(\Omega, \mathcal{F}, \mu)$. Then the set $\mathcal{S}_\Gamma + \mathcal{H}$ is convex and closed in measure.*

Proof. We need to prove that if (u_n) is a sequence in $\mathcal{S}_\Gamma + \mathcal{H}$ which converges almost everywhere to a function $u \in \mathcal{L}_E^0(\Omega, \mathcal{F}, \mu)$, then $u \in \mathcal{S}_\Gamma + \mathcal{H}$. There are $f_n \in \mathcal{S}_\Gamma$ and $g_n \in \mathcal{H}$ such that $u_n = f_n + g_n$ for all n. Repeating the arguments given in the proof of Proposition 2.2, there exist a subsequence $(f_{\gamma(n)})$ of (f_n) and $f \in \mathcal{S}_\Gamma$ such that

$$\lim_{n\to\infty} \frac{1}{n} \sum_{j=1}^{n} f_{\gamma(j)}(\omega) = f(\omega)$$

for almost every $\omega \in \Omega$, with respect to the norm topology. Now it is obvious that the sequence $(\frac{1}{n} \sum_{j=1}^{n} g_{\gamma(j)})$ converges almost everywhere to $g := u - f$. Since $\mathcal{H}$ is convex closed in measure, we have $g \in \mathcal{H}$. Thus $u = f + g \in \mathcal{S}_\Gamma + \mathcal{H}$.

□

3. Some Minimization problems and min–max type results

We give some applications of the preceding results to some Minimization problems.

Proposition 3.1. *Let E be a separable super reflexive Banach space, let Γ be a convex weakly compact random set and let $\mathcal{S}_\Gamma$ be the set of all $\mathcal{F}$-measurable selections of Γ. Then $\mathcal{S}_\Gamma$ is convex and closed for the convergence in measure. Let $J : \mathcal{S}_\Gamma \to [0, \infty[$ be a convex and lower semicontinuous for the convergence in measure. Then J reaches its mimimum on $\mathcal{S}_\Gamma$.*

Proof. It is obvious that $\mathcal{S}_\Gamma$ is convex and closed for the convergence in measure. Let (u_n) be a minimizing sequence in $\mathcal{S}_\Gamma$, that is, $\lim_n J(u_n) = m := \inf\{J(u) : u \in \mathcal{S}_\Gamma\}$. Notice that $\sup_n ||u_n(\omega)| \mid \leq |\Gamma(\omega)| < +\infty$ for each $\omega \in \Omega$. In view of Theorem 2.1, there exists a subsequence $(u_{\gamma(n)})$ of (u_n) and $\overline{u} \in \mathcal{L}^0_E(\Omega, \mathcal{F}, \mu)$ such that

$$\lim_{n\to\infty} ||\frac{1}{n}\sum_{j=1}^{n} u_{\gamma(j)}(\omega) - \overline{u}(\omega)|| = 0$$

for almost every $\omega \in \Omega$. By convexity we have

$$J\left(\frac{1}{n}\sum_{j=1}^{n} u_{\gamma(j)}\right) \leq \Sigma_{j=1}^{n}\frac{1}{n}J(u_{\gamma(j)}).$$

Since J is lsc for the convergence in measure and $(\frac{1}{n}\sum_{j=1}^{n} u_{\gamma(j)})_n$ converges in measure to $\overline{u} \in \mathcal{S}_\Gamma$, we have

$$\liminf_{n\to\infty} J\left(\frac{1}{n}\sum_{j=1}^{n} u_{\gamma(j)}\right) \geq J(\overline{u}) \geq m.$$

But is obvious that

$$\lim_n \sum_{j=1}^{n}\frac{1}{n}J(u_{\gamma(j)}) = m.$$

It follows that

$$J(\overline{u}) = m := \inf\{J(u) : u \in \mathcal{S}_\Gamma\}.$$

□

There is a direct application which arises from evolution problems. See ([18], Theorem VII-18).

Proposition 3.2. *Let H be a separable Hilbert space, let dt be the Lebesgue measure on $[0, 1]$. Let $\Gamma : [0, 1] \Longrightarrow H$ be a convex weakly compact random set such that $|\Gamma| \in L^2_{\mathbf{R}}([0, 1], dt)$ and $\mathcal{S}^2_\Gamma$ the set of all $L^2_H([0, 1], dt)$-selections of Γ. For each $u \in \mathcal{S}^2_\Gamma$, let Pu be the primitive of $u \in L^2_H([0, 1], dt)$*

$$Pu(t) := \int_0^t u(s)ds, \quad \forall t \in [0, 1].$$

Let j be a nonnegative convex normal integrand defined on $[0, 1] \times H$ such that the associated convex integral functional

$$J(u) := \int_0^1 j(t, u(t))dt, \quad u \in L^2_H([0, 1], dt)$$

is finite on $\mathcal{S}^2_\Gamma$ and the conjugate convex integral functional

$$J^*(u) := \int_0^1 j^*(t, u(t))dt, \quad \forall u \in L^2_H([0, 1], dt)$$

is proper on $L^2_H([0, 1], dt)$. Let Ψ be the integral functional

$$\Psi(u) := \int_0^1 \langle Pu(t), u(t)\rangle dt + J(u), \quad \forall u \in L^2_H([0, 1], dt).$$

Then the functional $\Psi(u)$ reaches its minimum on $\mathcal{S}^2_\Gamma$.

Proof. (a) First proof. By a remark in ([18], Theorem VII-18), it is not difficult to check that

$$1/2 \int_0^1 \langle Pu(t), u(t)\rangle dt = \int_0^1 ||u(t)||^2 \, dt.$$

Using the duality of convex integral functional in ([18], Theorem VII-7), J is proper convex lower semicontinuous on $L^2_H([0, 1], dt)$ with respect to the $\sigma(L^2_H, L^2_H)$ topology, and is finite on $\mathcal{S}^2_\Gamma$, here the nonnegativity of j is unnecessary, so is Ψ. As it is obvious that $\mathcal{S}^2_\Gamma$ is nonempty convex and weakly compact in $L^2_H([0, 1], dt)$, it is immediate that Ψ reaches its minimum on $\mathcal{S}^2_\Gamma$.
(b) Second proof. It is obvious that $\mathcal{S}^2_\Gamma$ is nonempty convex and closed for the convergence in measure and the convex integral functional Ψ is finite, nonnegative, convex and lower semicontinuous on $\mathcal{S}^2_\Gamma$ for the convergence in measure. Then Proposition 3.1 shows that Ψ reaches its minimum on $\mathcal{S}^2_\Gamma$. □

Now we proceed to further results of Minimization with other techniques.

Proposition 3.3. *Let H be a separable Hilbert space, F be a nonempty weakly closed subset of $L^2_H([0,1],dt)$ and let $\mathcal{A} : F \to L^2_H([0,1],dt)$ be a compact mapping, (that is, $\mathcal{A}$ transforms any weakly convergent sequence in F into strongly convergent sequence in $L^2_H([0,1],dt)$), satisfying*

$$\langle \mathcal{A}u, u\rangle_{\langle L^2_H, L^2_H\rangle} \geq \alpha ||u||^2_{L^2_H}$$

for all $u \in F$, for some positive constant α. For each $u \in L^2_H([0,1],dt)$, let Pu be its primitive

$$Pu(t) := \int_0^t u(s)ds, \quad \forall t \in [0,1].$$

Then the functional integral $\Psi(u) := \langle \mathcal{A}u, u\rangle_{\langle L^2_H, L^2_H\rangle} + \int_0^1 \langle Pu(t), u(t)\rangle dt$ reaches its minimum on F.

Proof. It is immediate that

$$\Psi(u) = \langle \mathcal{A}u, u\rangle_{\langle L^2_H, L^2_H\rangle} + \frac{1}{2}||u||^2_{L^2_H}$$

verifies

$$\Psi(u) \geq \left(\alpha + \frac{1}{2}\right) ||u||^2_{L^2_H}$$

for all $u \in F$. Let (u_n) be a minimizing sequence for Ψ, that is $\lim_n \Psi(u_n) = \inf_{u\in F} \Psi(u)$ with $u_n \in F$ for all n. We may assume that

$$\Psi(u_n) \leq 1 + \inf_{u\in F} \Psi(u)$$

for all n, so that, using the preceding estimate, we get

$$||u_n||^2_{L^2_H} \leq \left(\alpha + \frac{1}{2}\right)^{-1} \left(1 + \inf_{u\in F} \Psi(u)\right)$$

for all n. Hence the sequence (u_n) is relatively weakly compact in $L^2_H([0,1],dt)$. So we may extract a subsequence (u_n) not relabelled which $\sigma(L^2_H, L^2_H)$ converges to $u \in F$ because F is weakly closed. As $J(u) := \frac{1}{2}||u||^2_{L^2_H}$ is lower semicontinuous on $L^2_H([0,1],dt)$ with respect to the topology $\sigma(L^2_H, L^2_H)$, it follows that

$$\begin{aligned}\inf_{u\in F} \Psi(u) = \lim_n \Psi(u_n) &= \lim_n \left[\langle \mathcal{A}u_n, u_n\rangle_{\langle L^2_H, L^2_H\rangle} + J(u_n)\right] \\ &\geq \langle \mathcal{A}u, u\rangle_{\langle L^2_H, L^2_H\rangle} + J(u).\end{aligned}$$

□

The following is a variant of the preceding ones.

Proposition 3.4. *Let F be a separable Banach space. Let G be a nonempty weakly compact subset of $L^1_F([0,1],dt)$ and let W be the closed unit ball of $L^\infty_{F'}([0,1],dt)$. Let $\mathcal{A} : G \to W$ a mapping transforming weakly convergent sequences in G into sequences in W converging in measure with respect to the norm topology of F'. Let j be a convex normal integrand defined on* $[0,1] \times F$ *satisfying*

$$0 \le j(t,x) \le \beta||x|| + \gamma$$

for all $(t,x) \in [0,1] \times F$ for some positive constant β and γ. Assume that there is $\overline{v} \in L^\infty_{F'}([0,1],dt)$ such that the conjugate integral functional $J^*(\overline{v}) := \int_0^1 j^*(t,\overline{v}(t))dt$ is finite. Then the functional integral $\Phi(u) := \langle \mathcal{A}u, u\rangle_{\langle L^\infty_{F'}, L^1_F\rangle} + J(u)$ reaches its minimum on G.

Proof. By our assumption it is immediate that

$$\Phi(u) := \langle \mathcal{A}u, u\rangle_{\langle L^\infty_{F'}, L^1_F\rangle} + J(u)$$

verifies

$$|\Phi(u)| \le (1+\beta)||u||_{L^1_F} + \gamma$$

for all $u \in G$. Let (u_n) be a minimizing sequence for Φ, that is $\lim_n \Phi(u_n) = \inf_{u\in G} \Phi(u)$ with $u_n \in G$ for all n. We may assume that (u_n) converges $\sigma(L^1_F, L^\infty_{F'})$ (alias weakly) in $L^1_F([0,1],dt)$ to $u \in G$. By our assumption, the sequence $(\mathcal{A}u_n)$ is bounded in $L^\infty_{F'}([0,1],dt)$ and converges in measure to $\mathcal{A}u$ with respect to the norm topology of F'. By virtue of Castaing [10], see also Grothendieck [8] for the one dimensional case, we conclude that

$$\lim_n \langle \mathcal{A}u_n, u_n\rangle_{\langle L^\infty_{F'}, L^1_F\rangle} = \langle \mathcal{A}u, u\rangle_{\langle L^\infty_{F'}, L^1_F\rangle}.$$

As $J(u) := \int_0^1 j(t,u(t))dt$ is convex lower semicontinuous on $L^1_F([0,1],dt)$, when $L^1_F([0,1],dt)$ is endowed with the topology $\sigma(L^1_F, L^\infty_{F'})$, see ([18], Theorem VII-7), we have

$$\begin{aligned}\lim_n \Phi(u_n) = \inf_{u\in G} \Phi(u) &= \lim_n [\langle \mathcal{A}u_n, u_n\rangle_{\langle L^\infty_{F'}, L^1_F\rangle} + J(u_n)] \\ &\ge \langle \mathcal{A}u, u\rangle_{\langle L^\infty_{F'}, L^1_F\rangle} + J(u).\end{aligned}$$

□

As an application of Proposition 3.1 we provide the following min–max type result.

Proposition 3.5. *Let E be a separable super reflexive Banach space, Γ be a convex weakly compact random set, $\mathcal{S}_\Gamma$ the set of all measurable selections of Γ and $\mathcal{K}$ be a compact space. Assume that $\Phi : \mathcal{S}_\Gamma \times \mathcal{K} \to \mathbf{R}^+$ is such that for every fixed $u \in \mathcal{S}_\Gamma$, $\Phi(u, .)$ is upper semicontinuous on $\mathcal{K}$ and for every fixed $v \in \mathcal{K}$, $\Phi(., v)$ is convex and lower semicontinuous for the convergence in measure on $\mathcal{S}_\Gamma$. Then there exists a pair $(\tilde{u}, \tilde{v}) \in \mathcal{S}_\Gamma \times \mathcal{K}$ such that*

$$\max_{v\in\mathcal{K}} \min_{u\in\mathcal{S}_\Gamma} \Phi(u, v) \leq \Phi(\widetilde{u}, \widetilde{v}) \leq \min_{u\in\mathcal{S}_\Gamma} \max_{v\in\mathcal{K}} \Phi(u, v).$$

Proof. Let us set

$$p(u) := \max_{v\in\mathcal{K}} \Phi(u, v), \quad \forall u \in \mathcal{S}_\Gamma$$
$$q(v) := \min_{u\in\mathcal{S}_\Gamma} \Phi(u, v), \quad \forall v \in \mathcal{K}$$

Then $p(.)$ is convex lower semicontinuous on $\mathcal{S}_\Gamma$ for the convergence in measure and $q(.)$ is upper semicontinuous on $\mathcal{K}$. By virtue of Proposition 3.1, there is $\widetilde{u} \in \mathcal{S}_\Gamma$ such that

$$p(\widetilde{u}) = \min_{u\in\mathcal{S}_\Gamma} p(u).$$

As $q(.)$ is upper semicontinuous on $\mathcal{K}$, there is $\widetilde{v} \in \mathcal{K}$ such that

$$q(\widetilde{v}) = \max_{v\in\mathcal{K}} q(v).$$

So we get

$$q(\widetilde{v}) \leq \Phi(\widetilde{u}, \widetilde{v}) \leq p(\widetilde{u}).$$

□

Other variants of Proposition 3.5 are available. Compare with Proposition 8.3.3 in [13] and the results stated below.

The following is a min–max result involving the use of Sion's theorem.

Proposition 3.6. *Let E be a separable reflexive Banach space, and Z be a compact metric space. Let $j : [0, 1] \times E \times Z$ be a non-negative normal integrand satisfying the following conditions*
(i) for each $(t, z) \in [0, 1] \times Z$, $j(t, ., z)$ is convex lower semicontinuous on E,
(ii) for each $(t, x) \in [0, 1] \times E$, $j(t, x, .)$ is continuous on Z,
(iii) there is a constant $c > 0$ such that $0 \leq j(t, x, z) \leq c(1 + ||x||)$ for all $(t, x, z) \in [0, 1] \times E \times Z$.
Let $\mathcal{R} := \mathcal{Y}([0, 1], \mathcal{M}^1_+(Z))$ be the space of Young measures associated with Z, that is, the set of all Lebesgue measurable mappings $\lambda : [0, 1] \to \mathcal{M}^1_+(Z)$;

$\mathcal{M}^1_+(Z)$ being the compact metrizable space of all probability Radon measures defined on Z, endowed with the vague topology and let $\mathcal{H}$ be a convex weakly compact subset of $L^1_E([0,1], dt)$. Let us consider the integral functional

$$J(u,\lambda) := \int_0^1 \left[\int_Z j(t, u(t), z)\lambda_t(dz)\right] dt; \quad \forall (u,\lambda) \in \mathcal{H} \times \mathcal{R}.$$

Then the following hold:
(a) there is a pair $(\widetilde{u}, \widetilde{\lambda}) \in \mathcal{H} \times \mathcal{R}$ such that

$$\max_{\lambda\in\mathcal{R}} \min_{u\in\mathcal{H}} J(u,\lambda) \leq J(\widetilde{u}, \widetilde{\lambda}) \leq \min_{u\in\mathcal{H}} \max_{v\in\mathcal{R}} J(u,\lambda).$$

(b)

$$\max_{\lambda\in\mathcal{R}} \min_{u\in\mathcal{H}} J(u,\lambda) = \min_{u\in\mathcal{H}} \max_{v\in\mathcal{R}} J(u,\lambda).$$

Proof. (a) follows the same line of the proof of the preceding result. Nevertheless this need a careful look. Let us set

$$p(u) := \max_{\lambda\in\mathcal{R}} J(u,\lambda), \quad \forall u \in \mathcal{H}$$

and

$$q(\lambda) := \min_{u\in\mathcal{H}} J(u,\lambda), \quad \forall \lambda \in \mathcal{R}.$$

For each $\lambda \in \mathcal{R}$ the convex integral functional

$$J(u,\lambda) := \int_0^1 \left[\int_Z j(t, u(t), z)\lambda_t(dz)\right] dt$$

is convex lower semicontinuous on $\mathcal{H}$, indeed, it is easy to see that

$$\hat{j}_\lambda(t,x) := \int_Z j(t,x,z)\lambda_t(dz)$$

is a nonnegative convex normal integrand on $[0, 1 \times E$ satisfying (iii), so that by ([13], Theorem 8.1.4 or Theorem 8.1.6) the associated integral functional

$$J(u,\lambda) = \int_0^1 \hat{j}_\lambda(t, u(t))dt$$

is (sequentially) lower semicontinuous on $\mathcal{H}$ with respect to the $\sigma(L^1_E, L^\infty_{E'})$ topology and so it reaches a minimum on the weakly compact set $\mathcal{H}$. Now we prove that $q(.)$ is upper semicontinuous on $\mathcal{R}$. Indeed, for each $u \in \mathcal{H}$, we have the estimate

$$0 \leq j(t, u(t), z) \leq c(1 + ||u(t)||)$$

Hence $(t, z) \to j(t, u(t), z)$ is a Carathéodory integrable integrand. Consequently, for each $u \in \mathcal{H}$, the functional

$$\lambda \to \int_0^1 \left[\int_Z j(t, u(t), z) \lambda_t(dz) \right] dt$$

is affine and continuous on (the convex compact) $\mathcal{R}$, taking account into the definition of stable convergence for Young measures [2, 13]. Then q is concave upper semicontinuous on $\mathcal{R}$, recalling that it is compact for the $\sigma(L^\infty_{C(Z)'}, L^1_{C(Z)})$-topology (alias stable topology). See [2,13,18] for details. As $\mathcal{H}$ is weakly compact, and p is weakly lower semicontinuous on $\mathcal{H}$, there is an element $\widetilde{u} \in \mathcal{H}$ such that

$$p(\widetilde{u}) = \min_{u \in \mathcal{R}} p(u).$$

As $q(.)$ is upper semicontinuous on $\mathcal{R}$, there is $\widetilde{\lambda} \in \mathcal{R}$ such that

$$q(\widetilde{\lambda}) = \max_{\lambda \in \mathcal{R}} q(\lambda).$$

So we get

$$q(\widetilde{\lambda}) \leq J(\widetilde{u}, \widetilde{\lambda}) \leq p(\widetilde{u})$$

which proves (a).
(b) follows by applying Sion's theorem [24] and the above considerations to the convex lsc-concave usc function J on the product $\mathcal{H} \times \mathcal{R}$. □

Remarks. Variants of Proposition 3.6 are available by combining new techniques for Young measures ([13], Theorem 8.1.6) and a general version of inf - sup theorem due to Moreau [22].

Acknowlegments. We wish to thank the referee for his careful reading and useful comments.

References

1. Amrani, A., Castaing, C., Valadier, M.: Méthodes de troncature appliquées à des problèmes de convergence faible ou forte dans L^1. Arch. Rational Mech. Anal. **117**, 167–191 (1992)

2. Balder, E.J.: New fundamentals of Young measure convergence. In: Calculus of variations and optimal control, Haifa (1998), Chapman Hall, Boca Raton 24–48 (2000)
3. Balder, E.J.: A general approach to lower semicontinuity result and lower closure in optimal control theory. SIAM J. Control Optim. 22, 570–598 (1984)
4. Balder, E.J., Hess, Ch.: Two generalizations of Komlós theorem with lower closure-type aplications. J. Convex Anal. **3**, 25–44 (1996)
5. Benabdellah, H., Castaing, C.: Weak compactness and convergences in $L^1_{E'}[E]$. Adv. Math. Econ. **3**, 1–44 (2001)
6. Beer, G.: Topologies on closed and closed convex sets. Kluwer, Dordrecht 1993
7. Bourgain, J.: The Komlós theorem for vector valued functions. Wrije Universiteit, Brussels, 1979/12
8. Grothendieck, A.: Espaces vectoriels topologiques. Publicacão da Sociedade de Matemática de Sao Paulo (1964)
9. Castaing, C.: Quelques résultats de convergence des suites adaptées. Sém. Anal. Convexe Montp. **17**, 1–24 (1989)
10. Castaing, C.: Topologie de la convergence uniforme sur les parties uniformément intégrables de L^1_E et théorèmes de compacité faible das certains espaces du type Kothe-Orlicz. Séminaire d' Analyse Convexe **10**, 1–27 (1980)
11. Castaing, C.: Weak compactness and convergences in Bochner and Pettis integration. Vietnam J. Math. **24**(3), 241–286 (1996)
12. Castaing, C., Ezzaki, F.: Convergences for convex weakly compact random sets in B-convex reflexive Banach spaces. Atti Sem. Mat. Fis. University of Modena, **XLVI**, 123–149 (1998)
13. Castaing, C., Raynaud de Fitte, P., Valadier, M.: Young Measures on Topological Spaces. With applications in control theory and probability theory. Kluwer Dordrecht 2004
14. Castaing, C., Guessous, M.: Convergences in $L^1_X(\mu)$. Adv. Math. Econ. **1**, 17–37 (1999)
15. Castaing, C., Hess, Ch., Saadoune, M.: Tightness conditions and Integrability of the sequential weak upper limit of a sequence of Multifunctions. Working paper 2005
16. Castaing, C., Hess, Ch., Saadoune, M.: On various versions of Fatou lemma. Working paper 2006
17. Castaing, C., Saadoune, M.: Dunford–Pettis-types theorem and convergences in set-valued integration, J. Nonlinear Convex Anal. **1**(1), 37–71 (2000)
18. Castaing, C., Valadier, M.: Convex Analysis and Measurable Multifunctions, Lectures Notes in Mathematics, Springer, Berlin, 580, 1977
19. Garling, D.J.H.: Subsequence principles for vector-valued random variables. Math. Proc. Cambridge Philos. Soc. **86**, 301–311 (1979)
20. Komlós, J.: A generalization of a problem of Steinhaus. Acta Math. Acad. Sci. Hungar **18**, 217–229 (1967)
21. Krupa, G.: Komlós theorem for unbounded random sets. Set-Valued Anal. **8**(3), 237–251 (2000)
22. Moreau, J.J.: Théorèmes “inf-sup”. C. R. Acad. Sci. Paris, **258**, 2720–2722 (1964)
23. Saadoune, M.: A new extension of Komlós theorem in infinite dimensions. Application: weak compactness in L^1_X. Portugaliae Math. **55**, 113–128 (1998)
24. Sion, M.: On general minimax theorems, Pacific J. Math. **8**, 171–176 (1958)

Adv. Math. Econ. 10, 31–49 (2007)

Advances in
MATHEMATICAL
ECONOMICS

Capital-labor substitution and indeterminacy in continuous-time two-sector models*

Jean-Philippe Garnier[1], **Kazuo Nishimura**[2] **and Alain Venditti**[3]

[1] Université de la Méditerranée, GREQAM, Marseille 13002, France
(e-mail: loopkill@club-internet.fr)
[2] Institute of Economic Research, Kyoto University, Kyoto 606-8501, Japan
(e-mail: nishimura@kier.kyoto-u.ac.jp)
[3] CNRS, GREQAM, Marseille 13002, France
(e-mail: venditti@ehess.univ-mrs.fr)

Received: August 8, 2006
Revised: November 24, 2006

JEL classification: C62, E32, O41

Mathematics Subject Classification (2000): 37C70, 37N40, 37C75, 39A11

Abstract. The aim of this paper is to discuss the role of the elasticity of capital-labor substitution on the local determinacy properties of the steady state in a two-sector economy with CES technologies and sector-specific externalities.

Key words: sector-specific externalities, constant returns, capital-labor substitution, indeterminacy.

1. Introduction

Recently, a large number of papers have established the fact that locally indeterminate equilibria and sunspots fluctuations easily arise within two-sector infinite-horizon growth models with sector specific external effects in production and linear preferences.[1] For instance, considering Cobb–Douglas technologies, Benhabib and Nishimura [2] prove within a continuous-time model that the existence of local indeterminacy is obtained if and only if there is a reversal of factor intensities between the private and social levels. The consumption

* We are grateful to T. Seegmuller, K. Shimomura and an anonymous referee for useful comments and suggestions.

[1] See Benhabib and Farmer [1].

good has indeed to be capital intensive from the private perspective but labor intensive from the social perspective.

When CES technologies are introduced, a much larger set of configurations is compatible with local indeterminacy. In particular, Leontief or linear technologies for the consumption good sector may be considered. Therefore, we will consider CES technologies in a continuous-time model and provide a complete analysis of the local determinacy properties of equilibria. We will show that, even with asymmetric factor substitutability properties, local indeterminacy is still based on a reversal of factor intensities between the private and social levels, what we call quasi factor intensity. We will also show that local indeterminacy is compatible with a Leontief technology in the consumption good sector, and that this result is preserved if the elasticity of capital-labor substitution in that sector is not too large.

The paper is organized as follows: § 2 presents the basic model with the production structure, the intertemporal equilibrium and the steady state. In § 3 we provide all the results on the existence of local indeterminacy depending on different configurations for the CES coefficients at the private and social levels. § 4 finally contains some concluding comments. All the proofs are gathered in a final Appendix.

2. The model

2.1. The production structure

We consider an economy producing a pure consumption good y_0 and a pure capital good y_1. Each good is assumed to be produced by capital x_{0j} and labor x_{1j}, $j = 0, 1$, through a CES technology which contains sector specific externalities. The representative firm in each industry indeed faces the following function:

$$y_j = \left(\beta_{0j} x_{0j}^{-\rho_j} + \beta_{1j} x_{1j}^{-\rho_j} + e_j(X_{0j}, X_{1j})\right)^{-1/\rho_j}, \quad j = 0, 1, \tag{1}$$

with $\beta_{ij} > 0$, $\rho_j > -1$ and $\sigma_j = 1/(1 + \rho_j) \geq 0$ the elasticity of capital-labor substitution. The positive externalities are equal to

$$e_j(X_{0j}, X_{1j}) = b_{0j} X_{0j}^{-\rho_j} + b_{1j} X_{1j}^{-\rho_j},$$

with $b_{ij} \geq 0$ and X_{ij} denoting the average use of input i in sector j. We assume that these economy-wide averages are taken as given by each individual firm. At the equilibrium, since all firms of sector j are identical, we have $X_{ij} = x_{ij}$ and we may define the *social production functions* as follows:

$$y_j = \left(\hat{\beta}_{0j} x_{0j}^{-\rho_j} + \hat{\beta}_{1j} x_{1j}^{-\rho_j}\right)^{-1/\rho_j} \tag{2}$$

with $\hat{\beta}_{ij} = \beta_{ij} + b_{ij}$. The returns to scale are therefore constant at the social level, and decreasing at the private level. We assume that in each sector $j = 0, 1$, $\hat{\beta}_{0j} + \hat{\beta}_{1j} = 1$ so that the production functions collapse to Cobb–Douglas in the particular case $\rho_j = 0$. Labor is normalized to one, i.e. $x_{00} + x_{01} = 1$, and the total stock of capital is given by $x_1 = x_{10} + x_{11}$.

Choosing the consumption good as the numeraire, i.e. $p_0 = 1$, a firm in each industry maximizes its profit given the output price p_1, the rental rate of capital w_1 and the wage rate w_0. Its profit is

$$\pi_j = p_j y_j - w_0 x_{0j} - w_1 x_{1j}.$$

The first-order conditions subject to the private technologies (1) are

$$p_j \beta_{ij} \left(y_j / x_{ij}\right)^{1+\rho_j} = w_i, \quad i, j = 0, 1. \tag{3}$$

From (3) we have

$$x_{ij}/y_j = \left(p_j \beta_{ij} / w_i\right)^{\frac{1}{1+\rho_j}} \equiv a_{ij}(w_i, p_j), \quad i, j = 0, 1. \tag{4}$$

We call a_{ij} the input coefficients from the *private* viewpoint. If the agents take account of externalities as endogenous variables in profit maximization, the first-order conditions subject to the social technologies (2) are

$$p_j \hat{\beta}_{ij} \left(y_j / x_{ij}\right)^{1+\rho_j} = w_i, \quad i, j = 0, 1.$$

and the input coefficients become

$$\bar{a}_{ij}(w_i, p_j) = \left(p_j \hat{\beta}_{ij} / w_i\right)^{1/(1+\rho_j)}, \quad i, j = 0, 1. \tag{5}$$

We call $\bar{a}_{ij}$ the input coefficients from the *social* viewpoint. We also define

$$\hat{a}_{ij}(w_i, p_j) \equiv (\hat{\beta}_{ij}/\beta_{ij}) a_{ij}(w_i, p_j), \tag{6}$$

as the quasi-input coefficients from the *social* viewpoint, and it is easy to derive that

$$\hat{a}_{ij}(w_i, p_j) = \bar{a}_{ij}(w_i, p_j) \left(\hat{\beta}_{ij}/\beta_{ij}\right)^{\rho_j/(1+\rho_j)}.$$

Notice that $\hat{a}_{ij} = \bar{a}_{ij}$ if there is no externality coming from input i in sector j, i.e. $b_{ij} = 0$, or if the production function is Cobb–Douglas, i.e. $\rho_j = 0$. As we will show below, the factor-price frontier, which gives a relationship between input prices and output prices, is not expressed with the input coefficients from the social viewpoint but with quasi-input coefficients from the social viewpoint.

These quasi-input coefficients at the social level correspond to input coefficients evaluated at the equilibrium, i.e., taking into account that the sector-specific externalities are equal to the actual equilibrium amounts of input used in each sector. They differ from the input coefficients at the social level since the producers consider the external effects as given when they maximize profit.

Based on these input coefficients straightforward computations allow to establish various lemmas. We first show that the factor-price frontier is determined by the quasi-input coefficients from the social viewpoint.

Lemma 1. *Denote* $p = (1, p_1)'$, $w = (w_0, w_1)'$ *and* $\hat{A}(w, p) = [\hat{a}_{ij}(w_i, p_j)]$. *Then* $p = \hat{A}'(w, p)w$.

The factor market clearing equation depends on the input coefficients from the private perspective.

Lemma 2. *Denote* $x = (1, x_1)'$, $y = (y_0, y_1)'$ *and* $A(w, p) = [a_{ij}(w_i, p_j)]$. *Then* $A(w, p)y = x$.

The basic intuition for these two results is the following: since the producers consider the external effects as given when they maximize profit, the input demand functions do not directly depend on the externalities so that the factor market clearing equation is only determined from the input coefficients at the private level. On the contrary, since external effects actually modify the equilibrium amount of output in both sectors, the prices of the final goods are affected so that the factor-price frontier is determined from the quasi-input coefficients at the social level.

Note that at the equilibrium the rental rate is a function of the output price only, $w_1 = w_1(p_1)$, while outputs are functions of the capital stock and the output price, $y_j = y_j(x_1, p_1)$, $j = 0, 1$.

We now examine some comparative statics. The factor-price frontier satisfies the Stolper–Samuelson theorem:

Lemma 3. $dw_1/dp_1 = \dfrac{\hat{a}_{00}}{\hat{a}_{11}\hat{a}_{00} - \hat{a}_{10}\hat{a}_{01}}$.

The factor market clearing equation finally satisfies the Rybczynski theorem:

Lemma 4. $dy_1/dx_1 = \dfrac{a_{00}}{a_{11}a_{00} - a_{10}a_{01}}$.

In an optimal growth model without external effects, i.e. $b_{ij} = 0$, we have $\hat{A}(w, p) = A(w, p)$. The Rybczynski and Stolper–Samuelson theorems are equivalent since $[\partial y/\partial x] = [\partial w/\partial p]'$. However, in presence of externalities, the Rybczynski effects depend on the input coefficients from the private perspective while the Stolper–Samuelson effects depend on the quasi-input coefficients from the social perspective. The duality between these two effects is thus destroyed. This property is again explained by the same intuition as the one given above. Local indeterminacy of equilibria will be a consequence of this broken duality.

2.2. Intertemporal equilibrium and steady state

A representative agent optimizes a linear additively separable utility function with discount rate $\delta > 0$. This problem can be described as:

$$\max_{\{x_{ij}(t)\}} \int_0^{+\infty} \left(\beta_{00}x_{00}(t)^{-\rho_0} + \beta_{10}x_{10}(t)^{-\rho_0} + e_0(X_{00}(t), X_{10}(t))\right)^{-\frac{1}{\rho_0}} e^{-\delta t}dt$$

$$\begin{aligned}
\text{s.t.}\quad & y_1(t) = \left(\beta_{01}x_{01}(t)^{-\rho_1} + \beta_{11}x_{11}(t)^{-\rho_1} + e_1(X_{01}(t), X_{11}(t))\right)^{-\frac{1}{\rho_1}} \\
& \dot{x}_1(t) = y_1(t) - gx_1(t) \\
& 1 = x_{00}(t) + x_{01}(t) \\
& x_1(t) = x_{10}(t) + x_{11}(t) \\
& x_1(0) \text{ and } \{e_j(X_{0j}(t), X_{1j}(t))\}_{t\geq 0}, \quad j = 0, 1, \text{ given}
\end{aligned}$$

where $g > 0$ is the depreciation rate of the capital stock. We can write the modified Hamiltonian in current value as

$$\begin{aligned}
\mathcal{H} \;=\; & \left(\beta_{00}x_{00}(t)^{-\rho_0} + \beta_{10}x_{10}(t)^{-\rho_0} + e_0(X_{00}(t), X_{10}(t))\right)^{-\frac{1}{\rho_0}} \\
& + w_0(t)\,(1 - x_{00}(t) - x_{01}(t)) + w_1(t)\,(x_1(t) - x_{10}(t) - x_{11}(t)) \\
& + p_1(t)\left(\left(\beta_{01}x_{01}(t)^{-\rho_1} + \beta_{11}x_{11}(t)^{-\rho_1} + e_1(X_{01}(t), X_{11}(t))\right)^{-\frac{1}{\rho_1}} - gx_1(t)\right).
\end{aligned}$$

The static first-order conditions are given by equation (3). The necessary conditions which describe the solution to the optimization problem are given by the following equations of motion:

$$\begin{aligned}
\dot{x}_1(t) &= y_1(x_1(t), p_1(t)) - gx_1(t) \\
\dot{p}_1(t) &= (\delta + g)p_1(t) - w_1(p_1(t)).
\end{aligned} \tag{7}$$

Any solution $\{x_1(t), p_1(t)\}_{t\geq 0}$ that also satisfies the transversality condition

$$\lim_{t\to+\infty} e^{-\delta t} p_1(t)x_1(t) = 0$$

is called an equilibrium path.

A steady state is defined by a pair (x_1^*, p_1^*) solution of

$$\begin{aligned}
y_1(x_1, p_1) &= gx_1 \\
w_1(p_1) &= (\delta + g)p_1.
\end{aligned} \tag{8}$$

We introduce the following restriction on parameters' values which will ensure the existence of an unique non trivial steady state:

Assumption 1. $\beta_{11} > \delta + g$ and $\rho_1 \in (\hat{\rho}_1, +\infty)$ with

$$\hat{\rho}_1 \equiv \frac{ln\hat{\beta}_{11}}{ln\left(\frac{\beta_{11}}{\delta+g}\right) - ln\hat{\beta}_{11}} \in (-1, 0) \tag{9}$$

Considering the fact that, within continuous-time models, the discount rate δ and the capital depreciation rate g are quite small, the restriction $\beta_{11} > \delta+g$ does not appear to be too demanding. Assumption 1 precisely guarantees positiveness and interiority of all the steady state values for input demand functions x_{ij}. Moreover, it allows to prove the following result:

Proposition 1. *Under Assumption 1, there exists a unique steady state* $(x_1^*, p_1^*) > 0$ *such that*

$$x_1^* = \frac{\left(\frac{\beta_{10}\beta_{01}}{\beta_{00}\beta_{11}}\right)^{\frac{1}{1+\rho_0}} \left(\frac{\left(\frac{\beta_{11}}{\delta+g}\right)^{\frac{\rho_1}{1+\rho_1}} - \hat{\beta}_{11}}{\hat{\beta}_{01}}\right)^{\frac{1+\rho_1}{\rho_1(1+\rho_0)}}}{1 - \left(\frac{\beta_{11}}{\delta+g}\right)^{\frac{1}{1+\rho_1}} g \left[1 - \left(\frac{\beta_{10}\beta_{01}}{\beta_{00}\beta_{11}}\right)^{\frac{1}{1+\rho_0}} \left(\frac{\left(\frac{\beta_{11}}{\delta+g}\right)^{\frac{\rho_1}{1+\rho_1}} - \hat{\beta}_{11}}{\hat{\beta}_{01}}\right)^{\frac{\rho_1-\rho_0}{\rho_1(1+\rho_0)}}\right]}$$

$$p_1^* = \frac{\beta_{10}}{\delta+g}\left[\hat{\beta}_{00}\left(\frac{\beta_{10}\beta_{01}}{\beta_{00}\beta_{11}}\right)^{\frac{\rho_0}{1+\rho_0}} \left(\frac{\left(\frac{\beta_{11}}{\delta+g}\right)^{\frac{\rho_1}{1+\rho_1}} - \hat{\beta}_{11}}{\hat{\beta}_{01}}\right)^{\frac{\rho_0(1+\rho_1)}{\rho_1(1+\rho_0)}} + \hat{\beta}_{10}\right]^{-\frac{1+\rho_0}{\rho_0}}.$$

3. Main results

We start by linearizing the dynamical system (7) around (x_1^*, p_1^*):

$$J = \begin{pmatrix} \frac{\partial y_1}{\partial x_1}(x_1^*, p_1^*) - g & \frac{\partial y_1}{\partial p_1}(x_1^*, p_1^*) \\ 0 & -\frac{\partial w_1}{\partial p_1}(p_1^*) + \delta + g \end{pmatrix}.$$

Any solution from (7) that converges to the steady state (x_1^*, p_1^*) satisfies the transversality condition and is an equilibrium. Therefore, given $x_1(0)$, if there is more than one initial price $p_1(0)$ in the stable manifold of (x_1^*, p_1^*), the equilibrium path from $x_1(0)$ will not be unique. In particular, if J has two roots with negative real parts, there will be a continuum of converging paths and thus a continuum of equilibria.

Definition 1. If the locally stable manifold of the steady state (x_1^*, p_1^*) is two-dimensional, then (x_1^*, p_1^*) is said to be locally indeterminate.

The roots of J are given by the diagonal terms. We know from Lemmas 3 and 4 that $\partial y_1/\partial x_1$ corresponds to the factor intensity difference from the private viewpoint and $\partial w_1/\partial p_1$ corresponds to the quasi-factor intensity difference from the social viewpoint. Using the definitions of input coefficients given in § 2, we may indeed interpret the elements of $\partial y_1/\partial x_1$ and $\partial w_1/\partial p_1$ as follows:

Definition 2. The consumption good is said to be:

(i) capital intensive at the **private** level if and only if $a_{11}a_{00} - a_{10}a_{01} < 0$,
(ii) **quasi** capital intensive at the **social** level if and only if $\hat{a}_{11}\hat{a}_{00} - \hat{a}_{10}\hat{a}_{01} < 0$,
(iii) capital intensive at the **social** level if and only if $\bar{a}_{11}\bar{a}_{00} - \bar{a}_{10}\bar{a}_{01} < 0$.

We may thus relate the input coefficients to the CES parameters:

Proposition 2. *Let Assumption 1 hold. At the steady state:*

(i) *the consumption good is capital (labor) intensive from the private perspective if and only if*

$$b \equiv 1 - \left(\frac{\beta_{10}\beta_{01}}{\beta_{00}\beta_{11}}\right)^{\frac{1}{1+\rho_0}} \left(\frac{\left(\frac{\beta_{11}}{\delta+g}\right)^{\frac{\rho_1}{1+\rho_1}} - \hat{\beta}_{11}}{\hat{\beta}_{01}}\right)^{\frac{\rho_1-\rho_0}{\rho_1(1+\rho_0)}} < (>)0$$

(ii) *the consumption good is quasi capital (labor) intensive from the social perspective if and only if*

$$\hat{b} \equiv 1 - \frac{\hat{\beta}_{10}\hat{\beta}_{01}}{\hat{\beta}_{00}\hat{\beta}_{11}} \left(\frac{\beta_{00}\beta_{11}}{\beta_{10}\beta_{01}}\right)^{\frac{\rho_0}{1+\rho_0}} \left(\frac{\left(\frac{\beta_{11}}{\delta+g}\right)^{\frac{\rho_1}{1+\rho_1}} - \hat{\beta}_{11}}{\hat{\beta}_{01}}\right)^{\frac{\rho_1-\rho_0}{\rho_1(1+\rho_0)}} < (>)0$$

(iii) *the consumption good is capital (labor) intensive from the social perspective if and only if*

$$\bar{b} \equiv 1 - \left(\frac{\hat{\beta}_{10}\hat{\beta}_{01}}{\hat{\beta}_{00}\hat{\beta}_{11}}\right)^{\frac{1}{1+\rho_0}} \left(\frac{\left(\frac{\hat{\beta}_{11}}{\delta+g}\right)^{\frac{\rho_1}{1+\rho_1}} - \hat{\beta}_{11}}{\hat{\beta}_{01}}\right)^{\frac{\rho_1-\rho_0}{\rho_1(1+\rho_0)}} < (>)0.$$

The following proposition establishes that local indeterminacy requires a capital intensity reversal from the private input coefficients to the quasi-input coefficients.

Proposition 3. *Let Assumption 1 hold. The steady state is locally indeterminate if and only if the consumption good is capital intensive from the private perspective, but quasi labor intensive from the social perspective.*

To get indeterminacy in a framework with constant returns to scale at the social level, we need a mechanism that nullify the duality between the Rybczynski and Stolper–Samuelson effects. As shown in § 2.1, the Rybczynski effect is given by the input coefficients from the private perspective while the Stolper–Samuelson effect is given by the quasi-input coefficients from the social perspective. In the presence of external effects, the duality between these coefficients is broken and local indeterminacy may appear.

This mechanism is very similar to the one exhibited in the contribution of Benhabib and Nishimura [2]. However in the current paper, it follows that depending on the value of the elasticities of capital-labor substitution, the capital intensity reversal from the private input coefficients to the quasi-input coefficients does not necessarily requires a capital intensity reversal from the private to the social level.

Propositions 2 and 3 show that the stability properties of the steady state depend, among all the parameters, on whether the ratios $\beta_{10}\beta_{01}/\beta_{11}\beta_{00}$ and $\hat{\beta}_{10}\hat{\beta}_{01}/\hat{\beta}_{11}\hat{\beta}_{00}$ are lower or greater than 1. Around the steady state, it can be easily shown that if the elasticities of capital-labor substitution are identical across sectors, the consumption good is capital intensive at the private level if and only if $\beta_{10}\beta_{01}/\beta_{11}\beta_{00} > 1$ while it is capital intensive at the social level if and only if $\hat{\beta}_{10}\hat{\beta}_{01}/\hat{\beta}_{11}\hat{\beta}_{00} > 1$. With asymmetric elasticities, as it is shown in Proposition 2, the capital intensity differences between sectors also depend on the parameters ρ_0 and ρ_1.

Notice also from Proposition 2 that the capital intensity differences at the private and quasi social level are linked as follows:

$$\hat{b} = 1 - \frac{\frac{\hat{\beta}_{10}\hat{\beta}_{01}}{\hat{\beta}_{11}\hat{\beta}_{00}}}{\frac{\beta_{10}\beta_{01}}{\beta_{11}\beta_{00}}}(1 - b).$$

We know from Proposition 3 that local indeterminacy requires the consumption good to be capital intensive at the private level, i.e. $b < 0$, but quasi labor intensive at the social level, i.e. $\hat{b} > 0$. It follows from the above expression that when $b < 0$, a necessary condition for $\hat{b}$ to be positive is given by the following Assumption:

Assumption 2. $\frac{\hat{\beta}_{10}\hat{\beta}_{01}}{\hat{\beta}_{11}\hat{\beta}_{00}} < \frac{\beta_{10}\beta_{01}}{\beta_{11}\beta_{00}}$.

Let us first consider the configuration $\beta_{10}\beta_{01}/\beta_{11}\beta_{00} < 1$ which is known in the case with symmetric elasticities of capital-labor substitution to imply local determinacy of the steady state.[2] The following Proposition shows on the contrary that as soon as the elasticities are sufficiently asymmetric, there is room for local indeterminacy.

Proposition 4. *Under Assumptions 1–2, let*

$$1 > \frac{\beta_{10}\beta_{01}}{\beta_{11}\beta_{00}} \text{ and } 1 > \frac{\hat{\beta}_{10}\hat{\beta}_{01}}{\hat{\beta}_{11}\hat{\beta}_{00}} > \frac{\hat{\beta}_{01}}{\frac{\beta_{11}}{\delta+g} - \hat{\beta}_{11}}. \tag{10}$$

Then there are $\underline{\rho}_0 \in (-1, 0)$, $\bar{\rho}_0 > 0$ *such that the following results hold:*

(i) *for any given* $\rho_0 \in (\underline{\rho}_0, \bar{\rho}_0)$, *there exist* $\underline{\rho}_1 > \max\{0, \rho_0\}$ *and* $\bar{\rho}_1 > \underline{\rho}_1$ *such that the steady state is locally indeterminate when* $\rho_1 \in (\underline{\rho}_1, \bar{\rho}_1)$;
(ii) *for any given* $\rho_0 > \bar{\rho}_0$, *there exists* $\underline{\rho}_1 > \rho_0$ *such that the steady state is locally indeterminate when* $\rho_1 > \underline{\rho}_1$.

Before giving interpretations of our results, we need to justify the part of condition (10) which concerns the CES coefficients at the social level: this inequality ensures the existence of the bounds $\bar{\rho}_0$ and $\underline{\rho}_0$, and the possible occurrence of local indeterminacy when both elasticities of capital labor substitution are close to zero.

All the various conditions for the existence of indeterminacy in Proposition 4 are based on the restriction $\rho_1 > \rho_0$ which implies a lower elasticity of capital-labor substitution in the investment good sector than in the consumption good sector. Since under Assumption 1 the elasticity of capital-labor substitution in the investment good sector is necessarily finite, it clearly appears that the same condition has to be considered for the investment good sector. In case (i), if the elasticity of capital-labor substitution in the consumption good sector is large enough, the extreme configuration of a Leontief technology in the investment good sector is also ruled out. On the contrary, if the factor substitutability of the consumption good sector is low enough, local indeterminacy becomes compatible with a Leontief technology in the investment good sector.

Let us now consider the configuration with $\beta_{10}\beta_{01}/\beta_{11}\beta_{00} > 1$ and $\hat{\beta}_{10}\hat{\beta}_{01}/\hat{\beta}_{11}\hat{\beta}_{00} < 1$ which is known to imply the existence of local indeterminacy in the case with Cobb–Douglas technologies in both sectors.[3] Notice that Assumption 2 is then necessarily satisfied.

[2] See Benhabib and Nishimura [2] for Cobb–Douglas economies and Nishimura and Venditti [3] for CES economies with symmetric elasticities of capital-labor substitution.

[3] See Benhabib and Nishimura [2].

Proposition 5. *Under Assumption 1, let*

$$\frac{\beta_{10}\beta_{01}}{\beta_{11}\beta_{00}} > 1 \text{ and } 1 > \frac{\hat{\beta}_{10}\hat{\beta}_{01}}{\hat{\beta}_{11}\hat{\beta}_{00}} > \frac{\hat{\beta}_{01}}{\frac{\beta_{11}}{\delta+g} - \hat{\beta}_{11}}. \tag{11}$$

Then there are $\underline{\rho}_0 \in (-1, 0)$ and $\bar{\rho}_0 > 0$ such that the following results hold:

(i) *for any given $\rho_0 \in (\underline{\rho}_0, \bar{\rho}_0)$, there exist $\underline{\rho}_1 \in (\hat{\rho}_1, 0)$ and $\bar{\rho}_1 > 0$ such that the steady state is locally indeterminate when $\rho_1 \in (\underline{\rho}_1, \bar{\rho}_1)$;*
(ii) *for any given $\rho_0 > \bar{\rho}_0$, there exists $\underline{\rho}_1 \in (0, \rho_0)$ such that the steady state is locally indeterminate when $\rho_1 > \underline{\rho}_1$.*

Proposition 5 covers the formulation with symmetric elasticities of capital-labor substitution across sectors previously analyzed in Benhabib and Nishimura [2] with Cobb–Douglas technologies and Nishimura and Venditti [3] with CES technologies having symmetric elasticities of capital-labor substitution. The part of condition (11) which concerns the CES coefficients at the social level is introduced to ensure the possibility of local indeterminacy when the consumption good technology is Cobb–Douglas while the investment good technology is Leontief [see case (ii)]. This explains why such a condition does not occur under the assumption of symmetric elasticities. However, since the plausible values of δ and g are close to zero, this condition does not imply a strong restriction on the CES coefficients.

Contrary to the previous Proposition, the existence of local indeterminacy does not require the elasticity of capital-labor substitution in the consumption good sector to be larger than the one in the investment good sector. The extreme configuration of an infinite factor substitutability in the consumption good sector is again ruled out. Moreover, as shown in case (ii), the occurrence of local indeterminacy with factors complementarity in the investment good sector is not compatible with a Cobb–Douglas technology in the consumption good sector and requires an elasticity of capital-labor substitution significantly lower than 1.

Let us finally consider the configuration with $\hat{\beta}_{10}\hat{\beta}_{01}/\hat{\beta}_{11}\hat{\beta}_{00} > 1$ which is known in the case with symmetric elasticities of capital-labor substitution to be compatible with the existence of local indeterminacy only when the common elasticity is lower than one, or $\rho_0 = \rho_1 = \rho > 0$.[4] In the next proposition, we show that in the asymmetric case, ρ_0 has to be positive but under Assumption 1, ρ_1 can be negative, equal to 1 or positive.

Proposition 6. *Under Assumptions 1–2, let*

$$\frac{\beta_{10}\beta_{01}}{\beta_{11}\beta_{00}} > 1 \text{ and } \frac{\hat{\beta}_{10}\hat{\beta}_{01}}{\hat{\beta}_{11}\hat{\beta}_{00}} > 1. \tag{12}$$

---

[4] See Nishimura and Venditti [3].

Then there are $\underline{\rho}_0 > 0$ *and* $\bar{\rho}_0 > \underline{\rho}_0$ *such that the following results hold:*

(i) *for any given* $\rho_0 \in (\underline{\rho}_0, \bar{\rho}_0)$*, there exist* $\underline{\rho}_1 \in (\hat{\rho}_1, \rho_0)$ *and* $\bar{\rho}_1 > 0$ *such that the steady state is locally indeterminate when* $\rho_1 \in (\underline{\rho}_1, \bar{\rho}_1)$*;*

(ii) *for any given* $\rho_0 > \bar{\rho}_0$*, there exists* $\underline{\rho}_1 > 0$ *such that the steady state is locally indeterminate when* $\rho_1 > \underline{\rho}_1$*.*

Proposition 6 confirms only part of the conclusions obtained by Nishimura and Venditti [3] under symmetric elasticities of capital-labor substitution across sectors: local indeterminacy requires some elasticity lower than one but only in the consumption good sector. Indeed, the existence of multiple equilibria does not rely on particular restrictions for the factor substitutability in the investment good sector. In particular, some elasticities greater than 1 are compatible with local indeterminacy in case (i). However, as in the previous Propositions, a Leontief technology for the investment good can be reached only if the elasticity in the consumption good sector is low enough.

4. Concluding comments

Within the framework of CES technologies, we precisely show how much the elasticities of capital-labor substitution can differ across the industries while leading to local indeterminacy of equilibria. For instance, given a value for the elasticity of substitution in the consumption good sector, the elasticity of substitution in the investment good sector can be arbitrarily large.

5. Appendix

5.1. Proof of Lemma 1

Substituting (4) into the social production functions (2) gives

$$y_j = \left(\hat{\beta}_{0j} y_j^{-\rho_j} \left(\frac{p_j \beta_{0j}}{w_0} \right)^{\frac{-\rho_j}{1+\rho_j}} + \hat{\beta}_{1j} y_j^{-\rho_j} \left(\frac{p_j \beta_{1j}}{w_1} \right)^{\frac{-\rho_j}{1+\rho_j}} \right)^{-1/\rho_j}.$$

It follows that

$$p_j^{\frac{\rho_j}{1+\rho_j}} = \left(\frac{\hat{\beta}_{0j}}{\beta_{0j}} \left(\frac{\beta_{0j}}{w_0} \right)^{\frac{1}{1+\rho_j}} w_0 + \frac{\hat{\beta}_{1j}}{\beta_{1j}} \left(\frac{\beta_{1j}}{w_1} \right)^{\frac{1}{1+\rho_j}} w_1 \right)^{-1/\rho_j}.$$

Multiplying both sides by $p_j^{1/(1+\rho_j)}$ then gives

$$p_j = \hat{a}_{0j} w_0 + \hat{a}_{1j} w_1.$$

The result follows considering that $p_0 = 1$. □

5.2. Proof of Lemma 2

By definition $x_{ij} = a_{ij} y_j$ and thus

$$x_i = a_{i0} y_0 + a_{i1} y_1.$$

The result follows considering that $x_0 = 1$. □

5.3. Proof of Lemma 3

The result immediately follows from the fact that the function $\hat{A}(w, p)$ is homogeneous of degree zero in w and p. □

5.4. Proof of Lemma 4

The result follows from a direct differentiation of $x_i = a_{i0} y_0 + a_{i1} y_1$ under the assumption that prices are constant. □

5.5. Proof of Proposition 1

From equation (3) considered at the steady state with $y_1 = g x_1$ and $w_1 = (\delta + g) p_1$, we get

$$x_{11} = \left(\frac{\beta_{11}}{\delta + g} \right)^{\frac{1}{1+\rho_1}} g x_1.$$

Using now the social production function (2) for the investment good we derive

$$x_{01} = \left(\frac{\left(\frac{\beta_{11}}{\delta+g} \right)^{\frac{\rho_1}{1+\rho_1}} - \hat{\beta}_{11}}{\hat{\beta}_{01}} \right)^{-\frac{1}{\rho_1}} \left(\frac{\beta_{11}}{\delta + g} \right)^{\frac{1}{1+\rho_1}} g x_1$$

and thus

$$\frac{x_{01}}{x_{11}} = \left(\frac{\left(\frac{\beta_{11}}{\delta+g} \right)^{\frac{\rho_1}{1+\rho_1}} - \hat{\beta}_{11}}{\hat{\beta}_{01}} \right)^{-\frac{1}{\rho_1}}. \tag{13}$$

Finally, we easily obtain from (3):

$$\begin{aligned} \frac{\beta_{10}\beta_{01}}{\beta_{00}\beta_{11}} &= \left(\frac{x_{01}}{x_{11}} \right)^{1+\rho_1} \left(\frac{x_{10}}{x_{00}} \right)^{1+\rho_0} \\ \Leftrightarrow \quad \frac{x_{10}}{x_{00}} &= \left(\frac{\beta_{10}\beta_{01}}{\beta_{00}\beta_{11}} \right)^{\frac{1}{1+\rho_0}} \left(\frac{\left(\frac{\beta_{11}}{\delta+g} \right)^{\frac{\rho_1}{1+\rho_1}} - \hat{\beta}_{11}}{\hat{\beta}_{01}} \right)^{\frac{1+\rho_1}{\rho_1(1+\rho_0)}}. \end{aligned} \tag{14}$$

Considering (13), (14) and the fact that $x_{00} + x_{01} = 1$, $x_1 = x_{10} + x_{11}$, we get the final expression of x_1^*. Equation (3) for $i = 1$ and $j = 0$ gives

$$w_1 = \beta_{10}\left(\hat{\beta}_{11}\left(\frac{x_{10}}{x_{00}}\right)^{\rho_0} + \hat{\beta}_{10}\right)^{-\frac{1+\rho_0}{\rho_0}}.$$

The final expression of p_1^* is then derived from (14) and the fact that $w_1 = (\delta + g)p_1$. □

5.6. Proof of Proposition 2

(i) From (4) we derive

$$a_{11}a_{00} - a_{10}a_{01} = a_{00}a_{01}\left(\frac{a_{11}}{a_{01}} - \frac{a_{10}}{a_{00}}\right) = a_{00}a_{01}\frac{x_{11}}{x_{01}}\left(1 - \frac{x_{10}x_{01}}{x_{00}x_{11}}\right)$$

and the result follows after substitution of (13) and (14) into the previous expression.

(ii) From (6) we derive

$$\begin{aligned}\hat{a}_{11}\hat{a}_{00} - \hat{a}_{10}\hat{a}_{01} &= \hat{a}_{00}\hat{a}_{01}\left(\frac{\hat{\beta}_{11}}{\hat{\beta}_{01}}\frac{\beta_{01}}{\beta_{11}}\frac{a_{11}}{a_{01}} - \frac{\hat{\beta}_{10}}{\hat{\beta}_{00}}\frac{\beta_{00}}{\beta_{10}}\frac{a_{10}}{a_{00}}\right)\\ &= a_{00}a_{01}\frac{\hat{\beta}_{11}}{\hat{\beta}_{01}}\frac{\beta_{01}}{\beta_{11}}\frac{x_{11}}{x_{01}}\left(1 - \frac{\hat{\beta}_{10}\hat{\beta}_{01}}{\hat{\beta}_{00}\hat{\beta}_{11}}\frac{\beta_{00}\beta_{11}}{\beta_{10}\beta_{01}}\frac{x_{10}x_{01}}{x_{00}x_{11}}\right)\end{aligned}$$

and the result follows after substitution of (13) and (14) into the previous expression.

(iii) The result is derived as in (i) but considering instead the input coefficients at the social level (5). □

5.7. Proof of Proposition 3

From Lemma 2, $x_1 = a_{10}y_0 + a_{11}y_1$. Moreover, at the steady state, $y_1 = gx_1$, and it follows

$$a_{10}y_0 + ga_{11}x_1 = x_1 \quad \Leftrightarrow \quad a_{10}y_0 = [1 - ga_{11}]x_1 > 0$$

Therefore,

$$\frac{\partial y_1}{\partial x_1} - g = \frac{a_{00}}{a_{11}a_{00} - a_{10}a_{01}} - g = \frac{a_{00}[1 - ga_{11}] + a_{10}a_{01}g}{a_{11}a_{00} - a_{10}a_{01}} < 0$$

if and only if $a_{11}a_{00} - a_{10}a_{01} < 0$. From Lemma 1, $p_1 = \hat{a}_{01}w_0 + \hat{a}_{11}w_1$. Moreover, at the steady state, $(\delta + g)p_1 = w_1$, and it follows

$$(\delta+g)\hat{a}_{01}w_0+(\delta+g)\hat{a}_{11}w_1=w_1 \quad \Leftrightarrow \quad (\delta+g)\hat{a}_{01}w_0 = [1-(\delta+g)\hat{a}_{11}]w_1>0$$

Therefore,

$$\begin{aligned}\delta+g-\frac{\partial w_1}{\partial p_1} &= \delta+g-\frac{\hat{a}_{00}}{\hat{a}_{11}\hat{a}_{00}-\hat{a}_{10}\hat{a}_{01}}\\ &= -\frac{[1-(\delta+g)\hat{a}_{11}]\hat{a}_{00}+(\delta+g)\hat{a}_{10}\hat{a}_{01}}{\hat{a}_{11}\hat{a}_{00}-\hat{a}_{10}\hat{a}_{01}}<0\end{aligned}$$

if and only if $\hat{a}_{11}\hat{a}_{00}-\hat{a}_{10}\hat{a}_{01}>0$. □

5.8. Proof of Proposition 4

Proposition 3 shows that local indeterminacy occurs if and only if $b<0$ and $\hat{b}>0$. Under Assumption 1, $\beta_{11}>\delta+g$ and $\rho_1\in(\hat{\rho}_1,+\infty)$, while $\rho_0>-1$ without particular restriction. We will then study the sign of b and $\hat{b}$ for different values of ρ_0 and ρ_1 over these intervals. Our strategy consists in considering a fixed value for ρ_0 and varying ρ_1 in order to find intervals of values in which local indeterminacy occurs. To simplify the analysis, we have to impose some restrictions on the parameters ρ_0 and ρ_1 such that the following function

$$g(\rho_1)=\left(\frac{\left(\frac{\beta_{11}}{\delta+g}\right)^{\frac{\rho_1}{1+\rho_1}}-\hat{\beta}_{11}}{\hat{\beta}_{01}}\right)^{\frac{\rho_1-\rho_0}{\rho_1(1+\rho_0)}} \tag{15}$$

is monotone increasing.

Lemma 5. *Under Assumption 1, there is a* $\tilde{\rho}_0\in(-1,0)$ *such that for any given* $\rho_0>\tilde{\rho}_0$ *there exists* $\tilde{\rho}_1\in[\hat{\rho}_1,0)$ *such that* $g(\rho_1)$ *is a monotone increasing function for all* $\rho_1>\tilde{\rho}_1$.

Proof Notice first that

$$\frac{\partial}{\partial\rho_1}\left(\frac{\rho_1}{1+\rho_1}\right)=\frac{1}{(1+\rho_1)^2}>0,\ \frac{\partial}{\partial\rho_1}\left(\frac{\rho_1-\rho_0}{\rho_1(1+\rho_0)}\right)=\frac{\rho_0}{\rho_1^2(1+\rho_0)}.$$

It follows that for any given $\rho_0\geq 0$, $g(\rho_1)$ is a monotone increasing function when $\rho_1\geq 0$. Consider now the case $\rho_0\in(-1,0)$. We can write $g(\rho_1)$ as follows:

$$g(\rho_1)=\exp\left\{\frac{\rho_1-\rho_0}{\rho_1(1+\rho_0)}ln\left(\frac{\exp\left[\frac{\rho_1}{1+\rho_1}ln\left(\frac{\beta_{11}}{\delta+g}\right)\right]-\hat{\beta}_{11}}{\hat{\beta}_{01}}\right)\right\}\equiv\exp\{f(\rho_1)\}.$$

If $f(\rho_1)$ is monotone increasing, so is $g(\rho_1)$. We easily compute

$$f'(\rho_1) = \frac{\rho_0(1+\rho_1)^2\left[\left(\frac{\beta_{11}}{\delta+g}\right)^{\frac{\rho_1}{1+\rho_1}} - \hat{\beta}_{11}\right] ln\left(\frac{\left(\frac{\beta_{11}}{\delta+g}\right)^{\frac{\rho_1}{1+\rho_1}} - \hat{\beta}_{11}}{\hat{\beta}_{01}}\right) + \rho_1(\rho_1-\rho_0)\left(\frac{\beta_{11}}{\delta+g}\right)^{\frac{\rho_1}{1+\rho_1}} ln\left(\frac{\beta_{11}}{\delta+g}\right)}{\rho_1^2(1+\rho_1)^2(1+\rho_0)\left[\left(\frac{\beta_{11}}{\delta+g}\right)^{\frac{\rho_1}{1+\rho_1}} - \hat{\beta}_{11}\right]}.$$

We have now to compute the sign of $\lim_{\rho_1\to 0} f'(\rho_1)$. A Taylor expansion of order two allows to show that $\lim_{\rho_1\to 0} f'(\rho_1) > 0$ for any $\rho_0 > \tilde{\rho}_0$ with

$$\tilde{\rho}_0 = -\left[1 + ln\left(\frac{\beta_{11}}{\delta+g}\right)\frac{\hat{\beta}_{11}}{2\hat{\beta}_{01}}\right]^{-1} \in (-1, \hat{\rho}_1)$$

Notice finally that $\lim_{\rho_1\to\hat{\rho}_1} f'(\rho_1) = -\infty$ and $\lim_{\rho_1\to\hat{\rho}_1} f(\rho_1) = +\infty$ if $\rho_0 < \hat{\rho}_1$, while $\lim_{\rho_1\to\hat{\rho}_1} f'(\rho_1) = +\infty$ and $\lim_{\rho_1\to\hat{\rho}_1} f(\rho_1) = -\infty$ if $\rho_0 > \hat{\rho}_1$. We then conclude from all this that for any given $\rho_0 > \tilde{\rho}_0$, there exists $\tilde{\rho}_1 \in [\hat{\rho}_1, 0)$ such that $g(\rho_1)$ is a monotone increasing function for all $\rho_1 > \tilde{\rho}_1$. □

It clearly appears from Proposition 2 that under the conditions of Lemma 5, b and $\hat{b}$ are monotone decreasing functions of ρ_1.

We may now prove Proposition 4: in the first part of the proof we consider positive values of ρ_1 only. Notice that L'Hôpital's rule gives

$$\lim_{\rho_1\to 0} g(\rho_1) = \left(\frac{\beta_{11}}{\delta+g}\right)^{\frac{-\rho_0}{(1+\rho_0)\hat{\beta}_{01}}}. \tag{16}$$

Consider the capital intensity difference at the private level b. The previous result implies

$$\lim_{\rho_1\to 0} b = 1 - \left(\frac{\beta_{10}\beta_{01}}{\beta_{00}\beta_{11}}\right)^{\frac{1}{1+\rho_0}}\left(\frac{\delta+g}{\beta_{11}}\right)^{\frac{\rho_0}{(1+\rho_0)\hat{\beta}_{01}}}. \tag{17}$$

Notice that the right-hand side of (17) is a monotone function of ρ_0. Under Assumption 2 and $\beta_{10}\beta_{01}/\beta_{11}\beta_{00} < 1$, we derive $\lim_{\rho_1\to 0} b\big|_{\rho_0\geq 0} > 0$. It follows that there exists $\rho_0^1 \in [-1, 0)$ such that $\lim_{\rho_1\to 0} b > 0$ for any $\rho_0 > \rho_0^1$. Similarly, we get

$$\lim_{\rho_1\to +\infty} b = 1 - \left(\frac{\beta_{10}\beta_{01}}{\beta_{00}\beta_{11}}\right)^{\frac{1}{1+\rho_0}}\left(\frac{\frac{\beta_{11}}{\delta+g} - \hat{\beta}_{11}}{\hat{\beta}_{01}}\right)^{\frac{1}{1+\rho_0}}. \tag{18}$$

Notice again that the right-hand side of (18) is a monotone function of ρ_0. If the following condition holds

$$\frac{\beta_{10}\beta_{01}}{\beta_{00}\beta_{11}} \frac{\frac{\beta_{11}}{\delta+g} - \hat{\beta}_{11}}{\hat{\beta}_{01}} > 1, \tag{19}$$

then $\lim_{\rho_1 \to +\infty} b\big|_{\rho_0 \to +\infty} = 0_-$ and $\lim_{\rho_1 \to +\infty} b\big|_{\rho_0 \leq 0} < 0$. It follows that under (19), $\lim_{\rho_1 \to +\infty} b < 0$ for any $\rho_0 > -1$. Notice finally that if $\rho_0 = \rho_1$ we have $b > 0$. Considering Lemma 5 under (19), we then conclude that for any given $\rho_0 > \rho_0^1$, there exists $\rho_1^1 > \max\{0, \rho_0\}$ such that $b < 0$ for all $\rho_1 > \rho_1^1$.

Consider now the quasi capital intensity difference at the social level $\hat{b}$. equation (16) implies

$$\lim_{\rho_1 \to 0} \hat{b} = 1 - \frac{\frac{\hat{\beta}_{10}\hat{\beta}_{01}}{\hat{\beta}_{11}\hat{\beta}_{00}}}{\frac{\beta_{10}\beta_{01}}{\beta_{11}\beta_{00}}} \left(\frac{\beta_{10}\beta_{01}}{\beta_{00}\beta_{11}}\right)^{\frac{1}{1+\rho_0}} \left(\frac{\delta+g}{\beta_{11}}\right)^{\frac{\rho_0}{(1+\rho_0)\hat{\beta}_{01}}}. \tag{20}$$

Notice that the right-hand side of (20) is a monotone function of ρ_0. Under Assumption 2 and $\hat{\beta}_{10}\hat{\beta}_{01}/\hat{\beta}_{11}\hat{\beta}_{00} < 1$, we derive $\lim_{\rho_1 \to 0} \hat{b}\big|_{\rho_0 \geq 0} > 0$ so that there exists $\rho_0^2 \in [-1, 0)$ such that $\lim_{\rho_1 \to 0} \hat{b} > 0$ for any $\rho_0 > \rho_0^2$. Since $\hat{b} > b$, we get when $\rho_0 = \rho_0^1$, $b\big|_{\rho_1 = 0} = 0$ while $\hat{b}\big|_{\rho_1 = 0} > 0$ so that $\rho_0^2 < \rho_0^1$. Similarly, we have

$$\lim_{\rho_1 \to +\infty} \hat{b} = 1 - \frac{\frac{\hat{\beta}_{10}\hat{\beta}_{01}}{\hat{\beta}_{11}\hat{\beta}_{00}}}{\frac{\beta_{10}\beta_{01}}{\beta_{11}\beta_{00}}} \left(\frac{\beta_{10}\beta_{01}}{\beta_{00}\beta_{11}}\right)^{\frac{1}{1+\rho_0}} \left(\frac{\frac{\beta_{11}}{\delta+g} - \hat{\beta}_{11}}{\hat{\beta}_{01}}\right)^{\frac{1}{1+\rho_0}}. \tag{21}$$

Notice again that the right-hand side of (21) is a monotone function of ρ_0. If the following condition holds

$$\frac{\hat{\beta}_{10}\hat{\beta}_{01}}{\hat{\beta}_{11}\hat{\beta}_{00}} \frac{\frac{\beta_{11}}{\delta+g} - \hat{\beta}_{11}}{\hat{\beta}_{01}} > 1, \tag{22}$$

we get with $\lim_{\rho_1 \to +\infty} \hat{b}\big|_{\rho_0 = 0} < 0$ and $\lim_{\rho_1 \to +\infty} \hat{b}\big|_{\rho_0 \to +\infty} > 0$. Therefore, under (22), there exists $\rho_0^3 > 0$ such that $\lim_{\rho_1 \to +\infty} \hat{b} > 0$ for any $\rho_0 > \rho_0^3$. Considering Lemma 5 under (22), we then conclude the following results:

– for any given $\rho_0 \in (\rho_0^2, \rho_0^3)$, there exists $\rho_1^2 > 0$ such that $\hat{b} > 0$ for all $\rho_1 \in [0, \rho_1^2)$. Notice that since $\hat{b} > b$, $\rho_1^2 > \rho_1^1$;
– for any given $\rho_0 > \rho_0^3$, $\hat{b} > 0$ for all $\rho_1 > \rho_0$.

Notice that under Assumption 2, (22) implies (19). Therefore, under (22), we conclude that local indeterminacy occurs in the following cases:

(i) for any given $\rho_0 \in (\max\{\rho_0^1, \rho_0^2\}, \rho_0^3)$ and $\rho_1 \in (\underline{\rho}_1, \bar{\rho}_1)$ with $\underline{\rho}_1 = \rho_1^1 > \max\{0, \rho_0\}$ and $\bar{\rho}_1 = \rho_1^2 > \underline{\rho}_1$;
(ii) for any given $\rho_0 > \rho_0^3$ and $\rho_1 > \underline{\rho}_1$ with $\underline{\rho}_1 = \rho_1^2 > \rho_0$. □

5.9. Proof of Proposition 5

We use the same kind of arguments as in Proposition 4. Consider first the capital intensity difference at the private level b. Under Assumption 2 and $\beta_{10}\beta_{01}/\beta_{11}\beta_{00} > 1$, we derive from (17) that $\lim_{\rho_1 \to 0} b\big|_{\rho_0 \leq 0} < 0$ while $\lim_{\rho_1 \to 0} b\big|_{\rho \to +\infty} > 0$. Therefore, there exists $\rho_0^1 > 0$ such that $\lim_{\rho_1 \to 0} b < 0$ for any $\rho_0 \in (-1, \rho_0^1)$. Similarly, we derive from (18) that $\lim_{\rho_1 \to +\infty} b < 0$ for any $\rho_0 > -1$. Notice also that

$$b|_{\rho_1=\rho_0} = 1 - \left(\frac{\beta_{10}\beta_{01}}{\beta_{00}\beta_{11}}\right)^{\frac{1}{1+\rho_0}} < 0. \tag{23}$$

Considering Lemma 5, we then conclude the following results:

(a) for any given $\rho_0 \in (-1, \rho_0^1)$, $b < 0$ for all $\rho_1 \geq 0$;
(b) for any given $\rho_0 > \rho_0^1$, there exists $\rho_1^1 \in (0, \rho_0)$ such that $b < 0$ for all $\rho_1 > \rho_1^1$.

Consider now the quasi capital intensity difference at the social level $\hat{b}$. Under Assumption 2 and $\hat{\beta}_{10}\hat{\beta}_{01}/\hat{\beta}_{11}\hat{\beta}_{00} < 1$, we derive from (20) that $\lim_{\rho_1 \to 0} \hat{b}\big|_{\rho_0 \geq 0} > 0$ while $\lim_{\rho_1 \to 0} \hat{b}\big|_{\rho_0 \to -1} < 0$. Therefore, there exists $\rho_0^2 \in (-1, 0)$ such that $\lim_{\rho_1 \to 0} \hat{b} > 0$ for any $\rho_0 > \rho_0^2$. Similarly, we derive from (21) that $\lim_{\rho_1 \to +\infty} \hat{b}\big|_{\rho_0 = 0} < 0$ if (22) holds, while $\lim_{\rho_1 \to +\infty} \hat{b}\big|_{\rho_0 \to +\infty} > 0$. Therefore, under (22), there exists $\rho_0^3 > 0$ such that $\lim_{\rho_1 \to +\infty} \hat{b} > 0$ for any $\rho_0 > \rho_0^3$. Notice also that

$$\hat{b}|_{\rho_1=\rho_0} = 1 - \frac{\hat{\beta}_{10}\hat{\beta}_{01}}{\hat{\beta}_{11}\hat{\beta}_{00}} \left(\frac{\beta_{00}\beta_{11}}{\beta_{10}\beta_{01}}\right)^{\frac{\rho_0}{1+\rho_0}} \tag{24}$$

is an increasing function of ρ_0 since $\beta_{00}\beta_{11}/\beta_{10}\beta_{01} < 1$. Considering that $\rho_0^2 \in (-1, 0)$ is defined as $\lim_{\rho_1 \to 0} \hat{b}\big|_{\rho_0 = \rho_0^2} = 0$, we derive from (20):

$$\hat{b}|_{\rho_1=\rho_0=\rho_0^2} = 1 - \left(\frac{\beta_{11}}{\delta + g}\right)^{\frac{\rho_0^2}{(1+\rho_0^2)\hat{\beta}_{01}}} > 0. \tag{25}$$

Therefore, $\hat{b}|_{\rho_1=\rho_0} > 0$ for any $\rho_0 \geq \rho_0^2$.

Considering Lemma 5 under (22), we then conclude the following results:

(c) for any given $\rho_0 \in (\rho_0^2, \rho_0^3)$, there exists $\rho_1^2 > \max\{0, \rho_0\}$ such that $\hat{b} > 0$ for all $\rho_1 \in [0, \rho_1^2)$;
(c) for any given $\rho_0 > \rho_0^3$, $\hat{b} > 0$ for all $\rho_1 \geq 0$.

Therefore, under (22), we conclude from points (a)-(e) that local indeterminacy occurs in the following cases:

(i) for any given $\rho_0 \in (\rho_0^2, \max\{\rho_0^1, \rho_0^3\})$ and $\rho_1 \in (\underline{\rho}_1, \bar{\rho}_1)$ with $\underline{\rho}_1 \in (\hat{\rho}_1, 0)$ and $\bar{\rho}_1 = \rho_1^2 > \max\{0, \rho_0\}$;
(ii) for any given $\rho_0 > \max\{\rho_0^1, \rho_0^3\}$ and $\rho_1 > \underline{\rho}_1$ with $\underline{\rho}_1 = \rho_1^1 > 0$. □

5.10. Proof of Proposition 6

We use the same kind of arguments as in Proposition 5. Consider first the capital intensity difference at the private level b. Under Assumption 2 and $\beta_{10}\beta_{01}/\beta_{11}\beta_{00} > 1$, we have already shown in the proof of Proposition 5 that there exists $\rho_0^1 > 0$ such that:

(a) for any given $\rho_0 \in (-1, \rho_0^1)$, $b < 0$ for all $\rho_1 \geq 0$;
(b) for any given $\rho_0 > \rho_0^1$, there exists $\rho_1^1 \in (0, \rho_0)$ such that $b < 0$ for all $\rho_1 > \rho_1^1$.

Consider now the quasi capital intensity difference at the social level $\hat{b}$. Under Assumption 2 and $\hat{\beta}_{10}\hat{\beta}_{01}/\hat{\beta}_{11}\hat{\beta}_{00} > 1$, we derive from (20) that $\lim_{\rho_1 \to 0} \hat{b}\big|_{\rho_0=0} < 0$ while $\lim_{\rho_1 \to 0} \hat{b}\big|_{\rho_0 \to +\infty} > 0$. Therefore, there exists $\rho_0^2 > 0$ such that $\lim_{\rho_1 \to 0} \hat{b} > 0$ for any $\rho_0 > \rho_0^2$. Since $\hat{b} > b$, we get $\rho_0^2 < \rho_0^1$. Similarly, we derive from (21) that $\lim_{\rho_1 \to +\infty} \hat{b}\big|_{\rho_0=0} < 0$ while $\lim_{\rho_1 \to +\infty} \hat{b}\big|_{\rho_0 \to +\infty} > 0$. Therefore, there exists $\rho_0^3 > 0$ such that $\lim_{\rho_1 \to +\infty} \hat{b} > 0$ for any $\rho_0 > \rho_0^3$.
Now notice from (20) and (21) the following property:

$$\lim_{\rho_1 \to +\infty} \hat{b}\big|_{\rho_0=\rho_0^2} = 1 - \left(\frac{\frac{\beta_{11}}{\delta+g} - \hat{\beta}_{11}}{\hat{\beta}_{01}} \right)^{\frac{1}{1+\rho_0^2}} \left(\frac{\beta_{11}}{\delta+g} \right)^{\frac{\rho_0^2}{(1+\rho_0^2)\hat{\beta}_{01}}} < 0.$$

Therefore, we get $\rho_0^2 < \rho_0^3$.
Consider finally equations (24) and (26). Since $\rho_0^2 > 0$ is such that $\lim_{\rho_1 \to 0} \hat{b}\big|_{\rho_0=\rho_0^2} = 0$, we derive from (20): $\hat{b}|_{\rho_1=\rho_0=\rho_0^2} < 0$. On the contrary, since $\rho_0^3 > 0$ is such that $\lim_{\rho_1 \to +\infty} \hat{b}\big|_{\rho_0=\rho_0^3} = 0$, we derive from (21):

$$\hat{b}|_{\rho_1=\rho_0=\rho_0^3} = 1 - \left(\frac{\hat{\beta}_{01}}{\frac{\beta_{11}}{\delta+g} - \hat{\beta}_{11}} \right)^{\frac{1}{1+\rho_0^3}} > 0. \quad (26)$$

Therefore, we get $\hat{b}|_{\rho_1=\rho_0} > 0$ for any $\rho_0 \geq \rho_0^3$.
Considering Lemma 5, we then conclude the following results:

(c) for any given $\rho_0 \in (\rho_0^2, \rho_0^3)$, there exists $\rho_1^2 \in (0, \rho_0]$ such that $\hat{b} > 0$ for all $\rho_1 \in [0, \rho_1^2)$;
(c) for any given $\rho_0 > \rho_0^3$, $\hat{b} > 0$ for all $\rho_1 \geq 0$.

Notice that since $\hat{b} > b$, we get $\rho_1^2 > \rho_1^1$.

Therefore, we conclude from points (a)–(d) that local indeterminacy occurs in the following cases:

(i) for any given $\rho_0 \in (\rho_0^2, \max\{\rho_0^1, \rho_0^3\})$ and $\rho_1 \in (\underline{\rho}_1, \bar{\rho}_1)$ with $\underline{\rho}_1 \in (\hat{\rho}_1, \rho_0)$ and $\bar{\rho}_1 = \rho_1^2 \in (0, \rho_0]$;

(ii) for any given $\rho_0 > \max\{\rho_0^1, \rho_0^3\}$ and $\rho_1 > \underline{\rho}_1$ with $\underline{\rho}_1 = \rho_1^1 \in (0, \rho_0)$.

□

References

1. Benhabib, J., Farmer, R.: Indeterminacy and sunspots in macroeconomics. In: Taylor, J.B., Woodford, M. (eds) Handbook of Macroeconomics. Amsterdam North-Holland pp. 387–448 (1999)
2. Benhabib, J., Nishimura, K.: Indeterminacy and sunspots with constant returns. J. Econ. Theory **81**, 58–96 (1998)
3. Nishimura, K., Venditti, A.: Indeterminacy and the role of factor substitutability. Macroecon. Dyn. **8**, 436–465 (2004)

Adv. Math. Econ. 10, 51–64 (2007)

Advances in MATHEMATICAL ECONOMICS

Weak and strong convergence theorems for new resolvents of maximal monotone operators in Banach spaces

Takanori Ibaraki[1] **and Wataru Takahashi**[2]

[1] Information Security Promotion Agency, Nagoya University, Furo-cho, Chikusa-ku, Nagoya, Aichi 464-8601, Japan
(e-mail: ibaraki@nagoya-u.jp)

[2] Department of Mathematical and Computing Sciences, Tokyo Institute of Technology, Oh-okayama, Meguro-ku, Tokyo 152-8552, Japan
(e-mail: wataru@is.titech.ac.jp)

Received: September 11, 2006
Revised: November 2, 2006

JEL classification: C61, C63, C02

Mathematics Subject Classification (2000): 49M05, 47H05, 47H09

Abstract. In this paper, we prove weak and strong convergence theorems for new resolvents of a maximal monotone operator in a Banach space which are connected with the proximal point algorithm of Rockafellar (SIAM J. Control. Optim. 14:877–898, 1976). Using these results, we consider the problem of finding minimizers of convex functions defined on Banach spaces.

Key words: Banach space, generalized nonexpansive retraction, proximal point algorithm, convex minimization problem

1. Introduction

Let H be a Hilbert space and let T be a maximal monotone operator from H to H. It is well known that many problems in nonlinear analysis and optimization can be formulated as follows: Find

$$u \in H \quad \text{such that } 0 \in Tu. \tag{1.1}$$

Such a $u \in H$ is called a zero point (or a zero) of T. A well-known method for solving (1.1) in a Hilbert space H is the proximal point algorithm: $x_1 \in H$ and

$$x_{n+1} = J_{r_n} x_n, \quad n = 1, 2, \dots, \tag{1.2}$$

where $\{r_n\} \subset (0, \infty)$ and $J_r = (I + rT)^{-1}$ for all $r > 0$. This algorithm was first introduced by Martinet [5]. In [19], Rockafellar proved that if $\liminf_{n\to\infty} r_n > 0$ and $T^{-1}0 \neq \emptyset$, then the sequence $\{x_n\}$ defined by (1.2) converges weakly to a solution of (1.1). Motivated by Rockafellar's result, Kamimura and Takahashi [6] proved the following two convergence theorems.

Theorem 1.1 ([6]). *Let H be a Hilbert space and let $T \subset H \times H$ be a maximal monotone operator. Let $J_r = (I + rT)^{-1}$ for all $r > 0$ and let $\{x_n\}$ be a sequence defined as follows: $x_1 = x \in H$ and*

$$x_{n+1} = \alpha_n x_n + (1 - \alpha_n) J_{r_n} x_n, \quad n = 1, 2, \dots,$$

where $\{\alpha_n\} \subset [0, 1]$ and $\{r_n\} \subset (0, \infty)$ satisfy

$$\limsup_{n\to\infty} \alpha_n < 1 \quad \text{and} \quad \liminf_{n\to\infty} r_n > 0.$$

If $T^{-1}0 \neq \emptyset$, then the sequence $\{x_n\}$ converges weakly to an element v of $T^{-1}0$, where $v = \lim_{n\to\infty} P_{T^{-1}0} x_n$ and $P_{T^{-1}0}$ is the metric projection of H onto $T^{-1}0$.

Theorem 1.2 ([6]). *Let H be a Hilbert space and let $T \subset H \times H$ be a maximal monotone operator. Let $J_r = (I + rT)^{-1}$ for all $r > 0$ and let $\{x_n\}$ be a sequence defined as follows: $x_1 = x \in H$ and*

$$x_{n+1} = \alpha_n x + (1 - \alpha_n) J_{r_n} x_n, \quad n = 1, 2, \dots,$$

where $\{\alpha_n\} \subset [0, 1]$ and $\{r_n\} \subset (0, \infty)$ satisfy $\lim_{n\to\infty} \alpha_n = 0$, $\sum_{n=1}^{\infty} \alpha_n = \infty$, and $\liminf_{n\to\infty} r_n = \infty$. If $T^{-1}0 \neq \emptyset$, then the sequence $\{x_n\}$ converges strongly to $P_{T^{-1}0}(x)$, where $P_{T^{-1}0}$ is the metric projection of H onto $T^{-1}0$.

These results were extended to more general Banach spaces; see Kamimura and Takahashi [7–9], Ohsawa and Takahashi [13], Kohsaka and Takahashi [11] and Kamimura *et al.* [10]. Recently, Ibaraki and Takahashi [4] found new resolvents in a Banach space which are connected with a maximal monotone operator.

Our purpose in this paper is to extend Kamimura–Takahashi's theorems to Banach spaces by using new resolvents of a maximal monotone operator. Such problems were posed in [22,23].

In § 3, we prove a weak convergence theorem which generalizes Theorem 1.1. In § 4, we obtain a strong convergence theorem which generalizes Theorem 1.2. Using these theorems, we consider the problem of finding minimizers of convex functions defined on Banach spaces.

2. Preliminaries

Let E be a real Banach space with its dual E^*. We write $x_n \rightharpoonup x_0$ to indicate that the sequence $\{x_n\}$ converges weakly to x_0. Similarly, $x_n \to x_0$ will symbolize the strong convergence. A Banach space E is said to be strictly convex if

$$\|x\| = \|y\| = 1, \quad x \neq y \Rightarrow \left\| \frac{x+y}{2} \right\| < 1.$$

Also, E is said to be uniformly convex if for each $\varepsilon \in (0, 2]$, there exists $\delta > 0$ such that

$$\|x\| = \|y\| = 1, \quad \|x - y\| \geq \varepsilon \Rightarrow \left\| \frac{x+y}{2} \right\| \leq 1 - \delta.$$

The normalized duality mapping J from E into E^* is defined by

$$J(x) := \left\{ x^* \in E^* : \langle x, x^* \rangle = \|x\|^2 = \|x^*\|^2 \right\}, \quad \forall x \in E.$$

An operator $T \subset E \times E^*$ with domain $D(T) = \{x \in E : Tx \neq \emptyset\}$ and range $R(T) = \cup\{Tx : x \in D(T)\}$ is said to be monotone if $\langle x - y, x^* - y^* \rangle \geq 0$ for any $(x, x^*), (y, y^*) \in T$. An operator T is said to be strictly monotone if $\langle x - y, x^* - y^* \rangle > 0$ for any $(x, x^*), (y, y^*) \in T$ $(x \neq y)$. A monotone operator T is said to be maximal if its graph $G(T) = \{(x, x^*) : x^* \in Tx\}$ is not properly contained in the graph of any other monotone operator. If T is maximal monotone, then the set $T^{-1}0 = \{u \in E : 0 \in Tu\}$ is closed and convex. If E is reflexive and strictly convex, then a monotone operator T is maximal if and only if $R(J + \lambda T) = E^*$ for each $\lambda > 0$ (see [18,21] for more details).

A Banach space E is said to be smooth if

$$\lim_{t \to 0} \frac{\|x + ty\| - \|x\|}{t} \tag{2.1}$$

exists for each $x, y \in \{z \in E : \|z\| = 1\} (=: S(E))$. In this case, the norm of E is said to be Gâteaux differentiable. The space E is said to have a uniformly Gâteaux differentiable norm if for each $y \in S(E)$, the limit (2.1) is attained uniformly for $x \in S(E)$. The norm of E is said to be Fréchet differentiable if for each $x \in S(E)$, the limit (2.1) is attained uniformly for $y \in S(E)$. The norm of E is said to be uniformly Fréchet differentiable (and E is said to be uniformly smooth) if the limit (2.1) is attained uniformly for $x, y \in S(E)$.

We also know the following properties (see [20] for details):

1. $Jx \neq \emptyset$ for each $x \in E$.
2. J is a monotone operator.
3. If E is strictly convex, then J is one to one, that is,
   $x \neq y \ \Rightarrow \ Jx \cap Jy = \emptyset$.

4. If E is reflexive, then J is a mapping of E onto E^*.
5. If E is smooth, then the duality mapping J is single valued.
6. If E has a Fréchet differentiable norm, then J is norm to norm continuous.
7. If E has a uniformly Gâteaux differentiable norm, then J is norm to weak* uniformly continuous on each bounded subset of E.
8. E is strictly convex if and only if J is a strictly monotone operator.
9. E is uniformly convex if and only if E^* is uniformly smooth.

Let E be a smooth Banach space and consider the following function studied in Alber [1] and Kamimura and Takahashi [9]:

$$V(x, y) = \|x\|^2 - 2\langle x, Jy\rangle + \|y\|^2,$$

for each $x, y \in E$. It is obvious from the definition of V that

$$(\|x\| - \|y\|)^2 \leq V(x, y) \leq (\|x\| + \|y\|)^2, \tag{2.2}$$

for each $x, y \in E$. We also know that

$$V(x, y) = V(x, z) + V(z, y) + 2\langle x - z, Jz - Jy\rangle, \tag{2.3}$$

for each $x, y, z \in E$(see [9]). The following lemma is well known.

Lemma 2.1 ([9]). *Let E be a smooth and uniformly convex Banach space and let $\{x_n\}$ and $\{y_n\}$ be sequences in E such that either $\{x_n\}$ or $\{y_n\}$ is bounded. If* $\lim_{n\to\infty} V(x_n, y_n) = 0$, *then* $\lim_{n\to\infty} \|x_n - y_n\| = 0$.

Let E be a smooth Banach space and let D be a nonempty closed convex subset of E. A mapping $R : D \to D$ is called generalized nonexpansive if $F(R) \neq \emptyset$ and $V(Rx, y) \leq V(x, y)$ for each $x \in D$ and $y \in F(R)$, where $F(R)$ is the set of fixed points of R. Let C be a nonempty closed subset of E. A mapping $R : E \to C$ is said to be sunny if

$$R(Rx + t(x - Rx)) = Rx, \quad \forall x \in E, \ \forall t \geq 0.$$

A mapping $R : E \to C$ is said to be a retraction if $Rx = x$, $\forall x \in C$. If E is smooth and strictly convex, then a sunny generalized nonexpansive retraction of E onto C is uniquely decided (see [3,4,14]). Then, if E be a smooth and strictly convex, a sunny generalized nonexpansive retraction of E onto C is denoted by R_C. Let C be a nonempty closed subset of a Banach space E. Then C is said to be a sunny generalized nonexpansive retract (resp. a generalized nonexpansive retract) of E if there exists a sunny generalized nonexpansive retraction (resp. a generalized nonexpansive retraction) of E onto C (see [3,4, 14] for more details). The set of fixed points of such a generalized nonexpansive retraction is C.

The following result was obtained in [4].

Lemma 2.2 ([4]). *Let C be a nonempty closed subset of a smooth and strictly convex Banach space E. Let R_C be a retraction of E onto C. Then R_C is sunny and generalized nonexpansive if and only if*

$$\langle x - R_C x, JR_C x - Jy \rangle \geq 0,$$

for each $x \in E$ and $y \in C$, where J is the duality mapping of E.

Let E be a reflexive, strictly convex, and smooth Banach space with its dual E^*. If a monotone operator $B \subset E^* \times E$ is maximal, then $E = R(I + rBJ)$ for all $r > 0$ (see Proposition 4.1 in [4]). So, for each $r > 0$ and $x \in E$, we can consider the set $J_r x = \{z \in E : x \in z + rBJz\}$. From [4], $J_r x$ consists of one point. We denote such a J_r by $(I + rBJ)^{-1}$. J_r is called a generalized resolvent of B (see [4] for more details).

The following two results were obtained in [4].

Lemma 2.3 ([4]). *Let E be a reflexive and strictly convex Banach space with a Fréchet differentiable norm and let $B \subset E^* \times E$ be a maximal monotone operator with $B^{-1}0 \neq \emptyset$. Then the following hold:*

1. $D(J_r) = E$ *for each* $r > 0$.
2. $(BJ)^{-1}0 = F(J_r)$ *for each* $r > 0$, *where* $F(J_r)$ *is the set of fixed points of* J_r.
3. $(BJ)^{-1}0$ *is closed.*
4. J_r *is generalized nonexpansive for each* $r > 0$.
5. *For* $r > 0$ *and* $x \in E$, $\frac{1}{r}(x - J_r x) \in BJJ_r x$.

Theorem 2.4 ([4]). *Let E be a uniformly convex Banach space with a Fréchet differentiable norm and let $B \subset E^* \times E$ be a maximal monotone operator with $B^{-1}0 \neq \emptyset$. Then the following hold:*

1. *For each* $x \in E$, $\lim_{r\to\infty} J_r x$ *exists and belongs to* $(BJ)^{-1}0$.
2. *If* $Rx := \lim_{r\to\infty} J_r x$ *for each* $x \in E$, *then* R *is a sunny generalized nonexpansive retraction of* E *onto* $(BJ)^{-1}0$.

3. Weak convergence theorem

In this section, we first start with the following lemma. Compare this lemma with the results in Kamimura and Takahashi [9], and Kohsaka and Takahashi [11].

Lemma 3.1. *Let E be a reflexive, strictly convex, and smooth Banach space, let $B \subset E^* \times E$ be a maximal monotone operator with $B^{-1}0 \neq \emptyset$, and $J_r = (I + rBJ)^{-1}$ for all $r > 0$. Then*

$$V(x, J_r x) + V(J_r x, u) \leq V(x, u),$$

for all $r > 0$, $u \in (BJ)^{-1}0$, and $x \in E$.

Proof. Let $r > 0$, $u \in (BJ)^{-1}0$, and $x \in E$ be given. By the monotonicity of B, (2.3) and Lemma 2.3 (5), we have

$$\begin{aligned} V(x, u) &= V(x, J_r x) + V(J_r x, u) + 2\langle x - J_r x, JJ_r x - Ju\rangle \\ &= V(x, J_r x) + V(J_r x, u) + 2r\left\langle \frac{x - J_r x}{r} - 0, JJ_r x - Ju\right\rangle \\ &\geq V(x, J_r x) + V(J_r x, u). \end{aligned}$$

□

Next we can prove the following weak convergence theorem, which is a generalization of Kamimura–Takahashi's weak convergence theorem (Theorem 1.1).

Theorem 3.2. *Let E be a smooth and uniformly convex Banach space whose duality mapping J is weakly sequentially continuous. Let $B \subset E^* \times E$ be a maximal monotone operator, let $J_r = (I + rBJ)^{-1}$ for all $r > 0$ and let $\{x_n\}$ be a sequence defined as follows: $x_1 = x \in E$ and*

$$x_{n+1} = \alpha_n x_n + (1 - \alpha_n) J_{r_n} x_n, \quad n = 1, 2, \ldots,$$

where $\{\alpha_n\} \subset [0, 1]$ and $\{r_n\} \subset (0, \infty)$ satisfy

$$\limsup_{n\to\infty} \alpha_n < 1 \quad \textit{and} \quad \liminf_{n\to\infty} r_n > 0.$$

If $B^{-1}0 \neq \emptyset$, then the sequence $\{x_n\}$ converges weakly to an element of $(BJ)^{-1}0$.

Proof. Note that $B^{-1}0 \neq \emptyset$ implies $(BJ)^{-1}0 \neq \emptyset$. In fact, if $u^* \in B^{-1}0$, we obtain $0 \in Bu^*$ and hence $0 \in BJJ^{-1}u^*$. So, we have $J^{-1}u^* \in (BJ)^{-1}0$.

Put $y_n = J_{r_n} x_n$ for all $n \in \mathbb{N}$ and let $z \in (BJ)^{-1}0$ be given. We first prove that $\{x_n\}$ is bounded. From Lemma 3.1 and the convexity of $\|\cdot\|^2$, we have

$$\begin{aligned} V(x_{n+1}, z) &= V(\alpha_n x_n + (1 - \alpha_n) y_n, z) \\ &\leq \alpha_n V(x_n, z) + (1 - \alpha_n) V(y_n, z) \\ &\leq \alpha_n V(x_n, z) + (1 - \alpha_n)\{V(x_n, z) - V(x_n, y_n)\} \\ &\leq \alpha_n V(x_n, z) + (1 - \alpha_n) V(x_n, z) \\ &= V(x_n, z), \end{aligned}$$

for all $n \in \mathbb{N}$. Hence, $\lim_{n\to\infty} V(x_n, z)$ exists. So, we have from (2.2) that the sequence $\{x_n\}$ is bounded and

$$(1-\alpha_n)V(x_n, y_n) \le V(x_n, z) - V(x_{n+1}, z),$$

for all $n \in \mathbb{N}$. Then it follows from $\limsup_{n\to\infty} \alpha_n < 1$ that

$$\lim_{n\to\infty} V(x_n, y_n) = 0. \tag{3.1}$$

We also know that the sequence $\{x_n\}$ is bounded. From (3.1) and Lemma 2.1, we have that

$$\lim_{n\to\infty} \|x_n - y_n\| = 0. \tag{3.2}$$

Since $\{x_n\}$ is bounded, we have a subsequence $\{x_{n_i}\}$ of $\{x_n\}$ such that $x_{n_i} \rightharpoonup v \in E$ as $i \to \infty$. Then it follows from (3.2) that $y_{n_i} \rightharpoonup v \in E$ as $i \to \infty$. On the other hand, from (3.2) and $\liminf_{n\to\infty} r_n > 0$, we have

$$\lim_{n\to\infty} \frac{\|x_n - y_n\|}{r_n} = 0.$$

If $(z^*, z) \in B$, then it holds from monotonicity of B that

$$\left\langle z - \frac{x_{n_i} - y_{n_i}}{r_{n_i}}, z^* - Jy_{n_i} \right\rangle \ge 0$$

for all $i \in \mathbb{N}$. Since J is weakly sequentially continuous, letting $i \to \infty$, we get $\langle z, z^* - Jv\rangle \ge 0$. Then, the maximality of B implies $Jv \in B^{-1}0$. That is, $v \in (BJ)^{-1}0$.

Let $\{x_{n_i}\}$ and $\{x_{n_j}\}$ be two subsequences of $\{x_n\}$ such that $x_{n_i} \rightharpoonup v_1$ and $x_{n_j} \rightharpoonup v_2$. As above, we have $v_1, v_2 \in (BJ)^{-1}0$. Put

$$a = \lim_{n\to\infty} (V(x_n, v_1) - V(x_n, v_2)).$$

Note that

$$V(x_n, v_1) - V(x_n, v_2) = 2\langle x_n, Jv_2 - Jv_1\rangle + \|v_1\|^2 - \|v_2\|^2, \quad n = 1, 2, \ldots.$$

From $x_{n_i} \rightharpoonup v_1$ and $x_{n_j} \rightharpoonup v_2$, we have

$$a = 2\langle v_1, Jv_2 - Jv_1\rangle + \|v_1\|^2 - \|v_2\|^2 \tag{3.3}$$

and

$$a = 2\langle v_2, Jv_2 - Jv_1\rangle + \|v_1\|^2 - \|v_2\|^2, \tag{3.4}$$

respectively. Combining (3.3) and (3.4), we obtain

$$\langle v_1 - v_2, Jv_1 - Jv_2\rangle = 0.$$

Since J is strictly monotone, it follows that $v_1 = v_2$. Therefore, $\{x_n\}$ converges weakly to an element of $(BJ)^{-1}0$. □

We know that the duality mappings J on ℓ^p, $1 < p < \infty$ and smooth finite dimensional Banach spaces are weakly sequentially continuous. However, we do not know whether Theorem 3.2 hold without assuming that J is weakly sequentially continuous.

4. Strong convergence theorem

Let E be a reflexive, strictly convex, and smooth Banach space and let J be the duality mapping from E into E^*. Then J^{-1} is also single valued, one-to-one, and surjective, and it is the duality mapping from E^* into E. We make use of the following mapping V_1 studied in Alber [1], and Kohsaka and Takahashi [11]:

$$V_1(x, x^*) = \|x\|^2 - 2\langle x, x^*\rangle + \|x^*\|^2, \tag{4.1}$$

for all $x \in E$ and $x^* \in E^*$. In other words, $V_1(x, x^*) = V(x, J^{-1}(x^*))$ for all $x \in E$ and $x^* \in E^*$. For each $x^* \in E^*$, the mapping $g(x) = V_1(x, x^*)$ for all $x \in E$ is a continuous and convex function from E into $\mathbb{R}$. As in Kohsaka and Takahashi [11], we can prove the following lemma.

Lemma 4.1. *Let E be a reflexive, strictly convex, and smooth Banach space and let V_1 be as in* (4.1). *Then*

$$V_1(x, x^*) + 2\langle y, Jx - x^*\rangle \le V_1(x + y, x^*)$$

for all $x, y \in E$ and $x^* \in E^*$.

Proof. Let $x^* \in E^*$ be given. Define $g(x) = V_1(x, x^*)$ and $f(x) = \|x\|^2$ for all $x \in E$. We have

$$\partial g(x) = \partial\left(f - 2\langle \cdot, x^*\rangle\right)(x) = 2Jx - 2x^*$$

for all $x \in E$. By the definition of ∂g, we have

$$g(x) + 2\langle y, Jx - x^*\rangle \le g(x + y),$$

that is,

$$V_1(x, x^*) + 2\langle y, Jx - x^*\rangle \le V_1(x + y, x^*)$$

for all $x, y \in E$. □

Next we can prove the following strong convergence theorem, which is a generalization of Kamimura–Takahashi's strong convergence theorem (Theorem 1.2).

Theorem 4.2. *Let E be a uniformly convex and uniformly smooth Banach space and let $B \subset E^* \times E$ be a maximal monotone operator. Let $J_r = (I + rBJ)^{-1}$ for all $r > 0$ and let $\{x_n\}$ be a sequence defined as follows: $x_1 = x \in E$ and*

$$x_{n+1} = \alpha_n x + (1 - \alpha_n) J_{r_n} x_n, \quad n = 1, 2, \dots,$$

where $\{\alpha_n\} \subset [0, 1]$ and $\{r_n\} \subset (0, \infty)$ satisfy $\lim_{n\to\infty} \alpha_n = 0$, $\sum_{n=1}^{\infty} \alpha_n = \infty$, and $\lim_{n\to\infty} r_n = \infty$. If $B^{-1}0 \neq \emptyset$, then the sequence $\{x_n\}$ converges strongly to $R_{(BJ)^{-1}0}(x)$, where $R_{(BJ)^{-1}0}$ is a sunny generalized nonexpansive retraction of E onto $(BJ)^{-1}0$.

Proof. Put $y_n = J_{r_n} x_n$ for all $n \in \mathbb{N}$. We denote a sunny generalized nonexpansive retraction $R_{(BJ)^{-1}0}$ of E onto $(BJ)^{-1}0$ by R. We first prove that $\{x_n\}$ is bounded. It is obvious that $V(x_1, Rx) \le V(x, Rx)$. Suppose that $V(x_n, Rx) \le V(x, Rx)$ for some $n \in \mathbb{N}$. Then from Lemma 3.1 and the convexity of $\|\cdot\|^2$, we have

$$\begin{aligned} V(x_{n+1}, Rx) &= V(\alpha_n x + (1 - \alpha_n) y_n, Rx) \\ &\le \alpha_n V(x, Rx) + (1 - \alpha_n) V(y_n, Rx) \\ &\le \alpha_n V(x, Rx) + (1 - \alpha_n) V(x_n, Rx) \\ &\le \alpha_n V(x, Rx) + (1 - \alpha_n) V(x, Rx) \\ &= V(x, Rx). \end{aligned}$$

Hence, by induction, we have $V(x_n, Rx) \le V(x, Rx)$ for all $n \in \mathbb{N}$. From (2.2), the sequence $\{x_n\}$ is bounded. From Lemma 3.1, we have that $V(y_n, Rx) = V(J_{r_n} x_n, Rx) \le V(x_n, Rx)$ for all $n \in \mathbb{N}$. So, $\{y_n\}$ is also bounded. We next prove

$$\limsup_{n\to\infty} \langle x - Rx, Jx_n - JRx \rangle \le 0. \tag{4.2}$$

Put $z_n = x_{n+1}$ for all $n \in \mathbb{N}$. Since $\{Jz_n\}$ is bounded, without loss of generality, we have a subsequence $\{Jz_{n_i}\}$ of $\{Jz_n\}$ such that

$$\lim_{i\to\infty} \langle x - Rx, Jz_{n_i} - JRx \rangle = \limsup_{n\to\infty} \langle x - Rx, Jz_n - JRx \rangle$$

and $\{Jz_{n_i}\}$ converges weakly to some $v^* \in E^*$. From the definition of $\{x_n\}$, we have

$$z_n - y_n = \alpha_n (x - y_n),$$

for all $n \in \mathbb{N}$. Since $\{y_n\}$ is bounded and $\lim_{n\to\infty} \alpha_n = 0$, we have

$$\lim_{n\to\infty} \|z_n - y_n\| = \lim_{n\to\infty} \alpha_n \|x - y_n\| = 0. \tag{4.3}$$

Since E has a uniformly Gâteaux differentiable norm, the duality mapping J is norm to weak* uniformly continuous on each bounded subset of E. Therefore, we obtain from (4.3) that

$$Jz_{n_i} - Jy_{n_i} \rightharpoonup 0 \quad \text{as } i \to \infty.$$

This implies $Jy_{n_i} \rightharpoonup v^*$ as $i \to \infty$. On the other hand, from $\lim_{n\to\infty} r_n = \infty$, we have

$$\lim_{n\to\infty} \frac{\|x_n - y_n\|}{r_n} = 0.$$

If $(z^*, z) \in B$, then it holds from the monotonicity of B that

$$\left\langle z - \frac{x_{n_i} - y_{n_i}}{r_{n_i}}, z^* - Jy_{n_i} \right\rangle \geq 0$$

for all $i \in \mathbb{N}$. Letting $i \to \infty$, we get $\langle z, z^* - v^* \rangle \geq 0$. Then, the maximality of B implies $v^* \in B^{-1}0$. Put $v = J^{-1}v^*$. Applying Lemma 2.2, we obtain

$$\begin{aligned} \limsup_{n\to\infty} \langle x - Rx, Jz_n - JRx \rangle &= \lim_{i\to\infty} \langle x - Rx, Jz_{n_i} - JRx \rangle \\ &= \langle x - Rx, v^* - JRx \rangle \\ &= \langle x - Rx, Jv - JRx \rangle \leq 0. \end{aligned}$$

Finally, we prove that $\lim_{n\to\infty} x_n = Rx$. Let $\varepsilon > 0$ be given. From (4.2), we have $m \in \mathbb{N}$ such that

$$\langle x - Rx, Jx_n - JRx \rangle \leq \varepsilon, \tag{4.4}$$

for all $n \geq m$. If $n \geq m$, then it holds from Lemma 4.1 and (4.4) that

$$\begin{aligned} V(x_{n+1}, Rx) &= V_1(x_{n+1}, JRx) \\ &= V_1(\alpha_n x + (1 - \alpha_n)y_n, JRx) \\ &\leq V_1(\alpha_n x + (1 - \alpha_n)y_n - \alpha_n(x - Rx), JRx) \\ &\quad - 2\langle -\alpha_n(x - Rx), Jx_{n+1} - JRx \rangle \\ &= V_1(\alpha_n Rx + (1 - \alpha_n)y_n, JRx) \\ &\quad + 2\alpha_n \langle x - Rx, Jx_{n+1} - JRx \rangle \\ &\leq V(\alpha_n Rx + (1 - \alpha_n)y_n, Rx) + 2\alpha_n\varepsilon \\ &\leq \alpha_n V(Rx, Rx) + (1 - \alpha_n)V(y_n, Rx) + 2\alpha_n\varepsilon. \end{aligned}$$

From $V(Rx, Rx) = 0$ and Lemma 3.1, we have

$$\begin{aligned} V(x_{n+1}, Rx) &\leq (1 - \alpha_n)V(x_n, Rx) + 2\alpha_n\varepsilon \\ &= 2\varepsilon\{1 - (1 - \alpha_n)\} + (1 - \alpha_n)V(x_n, Rx). \end{aligned}$$

Therefore, we have

$$
\begin{aligned}
&V(x_{n+1}, Rx) \\
&\leq 2\varepsilon\{1-(1-\alpha_n)\} \\
&\quad +(1-\alpha_n)[2\varepsilon\{1-(1-\alpha_{n-1})\}+(1-\alpha_{n-1})V(x_{n-1}, Rx)] \\
&= 2\varepsilon\{1-(1-\alpha_n)(1-\alpha_{n-1})\}+(1-\alpha_n)(1-\alpha_{n-1})V(x_{n-1}, Rx) \\
&\leq \cdots \leq 2\varepsilon\left\{1-\prod_{i=m}^{n}(1-\alpha_i)\right\}+\prod_{i=m}^{n}(1-\alpha_i)V(x_m, Rx)
\end{aligned}
$$

for all $n \geq m$. Since $\sum_{i=1}^{\infty}\alpha_i = \infty$, we have $\prod_{i=m}^{\infty}(1-\alpha_i) = 0$ (see Takahashi [21]). Hence, we have

$$
\begin{aligned}
&\limsup_{n\to\infty} V(x_n, Rx) \\
&= \limsup_{l\to\infty} V(x_{m+l+1}, Rx) \\
&\leq \limsup_{l\to\infty}\left[2\varepsilon\left\{1-\prod_{i=m}^{m+l}(1-\alpha_i)\right\}+\prod_{i=m}^{m+l}(1-\alpha_i)V(x_m, Rx)\right] = 2\varepsilon.
\end{aligned}
$$

This implies $\limsup_{n\to\infty} V(x_n, Rx) \leq 0$. Hence, we get

$$\lim_{n\to\infty} V(x_n, Rx) = 0.$$

Applying Lemma 2.1, we obtain

$$\lim_{n\to\infty} \|x_n - Rx\| = 0.$$

Therefore, $\{x_n\}$ converges strongly to $R_{(BJ)^{-1}0}(x)$. □

5. Applications

In this section, we study the problem of finding a minimizer of a proper lower semicontinuous convex function; see [2,16,21].

As a direct consequence of Theorem 3.2, we obtain the following result.

Corollary 5.1. *Let E be a smooth and uniformly convex Banach space whose duality mapping J is weakly sequentially continuous. Let $f^* : E^* \to (-\infty, \infty]$ be a proper lower semicontinuous convex function such that $(\partial f^*)^{-1}(0) \neq \emptyset$. Let $\{x_n\}$ be a sequence defined as follows: $x_1 = x \in E$ and*

$$
\begin{aligned}
y_n^* &= \operatorname*{argmin}_{y^*\in E^*}\left\{f^*(y^*)+\frac{1}{2r_n}\|y^*\|^2-\frac{1}{r_n}\langle x_n, y^*\rangle\right\}, \qquad (5.1)\\
x_{n+1} &= \alpha_n x_n + (1-\alpha_n)J^{-1}y_n^*, \quad n = 1, 2, \ldots,
\end{aligned}
$$

where $\{\alpha_n\} \subset [0, 1]$ *and* $\{r_n\} \subset (0, \infty)$ *satisfy*

$$\limsup_{n\to\infty} \alpha_n < 1 \quad \textit{and} \quad \liminf_{n\to\infty} r_n > 0.$$

Then the sequence $\{x_n\}$ *converges weakly to an element of* $(\partial f^* J)^{-1}0$.

Proof. By Rockafellar's theorem [15,17], the subdifferential mapping $\partial f^* \subset E^* \times E$ is maximal monotone. Fix $r > 0$, $z \in E$, and let J_r be the generalized resolvent of ∂f^*. Then we have

$$z \in J_r z + r\partial f^* J J_r z$$

and hence,

$$0 \in \partial f^* J J_r z + \frac{1}{r} J^{-1} J J_r z - \frac{1}{r} z = \partial \left(f^* + \frac{1}{2r} \| \cdot \|^2 - \frac{1}{r}\langle z, \cdot \rangle \right) J J_r z.$$

Thus, we have

$$J J_r z = \operatorname*{argmin}_{y^* \in E^*} \left\{ f^*(y^*) + \frac{1}{2r}\|y^*\|^2 - \frac{1}{r}\langle z, y^* \rangle \right\}.$$

Therefore, $J^{-1} y_n^* = J^{-1} J J_{r_n} x_n = J_{r_n} x_n$ for all $n \in \mathbb{N}$. By Theorem 3.2, $\{x_n\}$ converges weakly to an element of $(\partial f^* J)^{-1}0$. □

As in the proof of Corollary 5.1, we get the following result from Theorem 4.2.

Corollary 5.2. *Let* E *be a uniformly convex and uniformly smooth Banach space and let* $f^* : E^* \to (-\infty, \infty]$ *be a proper lower semicontinuous convex function such that* $(\partial f^*)^{-1}(0) \neq \emptyset$. *Let* $\{x_n\}$ *be a sequence defined as follows:* $x_1 = x \in E$ *and*

$$\begin{aligned} y_n^* &= \operatorname*{argmin}_{y^* \in E^*} \left\{ f^*(y^*) + \frac{1}{2r_n}\|y^*\|^2 - \frac{1}{r_n}\langle x_n, y^* \rangle \right\}, \\ x_{n+1} &= \alpha_n x + (1 - \alpha_n) J^{-1} y_n^*, \quad n = 1, 2, \ldots, \end{aligned} \tag{5.2}$$

where $\{\alpha_n\} \subset [0, 1]$ *and* $\{r_n\} \subset (0, \infty)$ *satisfy* $\lim_{n\to\infty} \alpha_n = 0$, $\sum_{n=1}^{\infty} \alpha_n = \infty$, *and* $\lim_{n\to\infty} r_n = \infty$. *Then the sequence* $\{x_n\}$ *converges strongly to* $R_{(\partial f^* J)^{-1}0}(x)$, *where* $R_{(\partial f^* J)^{-1}0}$ *is a sunny generalized nonexpansive retraction of* E *onto* $(\partial f^* J)^{-1}0$.

References

1. Alber, Y.I.: Metric and generalized projection operators in Banach spaces: properties and applications. Theory and Applications of Nonlinear Operators of Accretive and Monotone Type, pp.15–50. Dekker, New York 1996
2. Butnariu, D., Iusem, A.N.: Totally Convex Functions for Fixed Points Computation and Infinite Dimensional Optimization. Kluwer Academic, Dordecht 2000
3. Ibaraki, T., Takahashi, W.: Convergence theorems for new projections in Banach spaces (in Japanese). RIMS Kokyuroku, vol. 1484, pp. 150–160, 2006
4. Ibaraki, T., Takahashi, W.: A new projection and convergence theorems for the projections in Banach spaces. (to appear)
5. Martinet, B.: Régularsation d'inéquations variationnells par approximations successives (in French). Rev. Francaise Informat. Rech. Opér. **4**, 154–158 (1970)
6. Kamimura, S., Takahashi, W.: Approximating solutions of maximal monotone operators in Hilbert spaces. J. Approx. Theory **106**, 226–240 (2000)
7. Kamimura, S., Takahashi, W.: Iterative schemes for approximating solutions of accritve operators in Banach spaces. Sci. Math. **3**, 107–115 (2000)
8. Kamimura, S., Takahashi, W.: Weak and strong convergence of solutions to accretive operator inclusions and applications. Set- Valued Anal. **8**, 361–374 (2000)
9. Kamimura, S., Takahashi, W.: Strong convergence of a proximal-type algorithm in a Banach space. SIAM J. Optim. **13**, 938–945 (2002)
10. Kamimura, S., Kohsaka, F., Takahashi, W.: Weak and strong convergence theorems for maximal monotone operators in a Banach space. Set Valued Anal. **12**, 417–429 (2004)
11. Kohsaka, F., Takahashi, W.: Strong convergence of an iterative sequence for maximal monotone operators in a Banach space. Abstr. Appl. Anal. **2004**, 239–249 (2004)
12. Matsushita, S., Takahashi, W.: Weak and strong convergence theorems for relatively nonexpansive mappings in Banach space. Fixed Point Theory Appl. **2004**, 37–47 (2004)
13. Ohsawa, S., Takahashi, W.: Strong convergence theorems for resolvents of maximal monotone operators in Banach spaces. Arch. Math. **81**, 439–445 (2003)
14. Reich, S.: Strong convergence theorems for resolvents of accretive operators in Banach space. J. Math. Anal. Appl. **75**, 287–292 (1980)
15. Rockafellar, R.T.: Characterization of the subdifferentials of convex functions. Pacific J. Math. **17**, 497–510 (1966)
16. Rockafellar, R.T.: Extension of Fenchel's duality theorem for convex functions. Duke Math. **33**, 81–89 (1966)
17. Rockafellar, R.T.: On the maximal monotonicity of subdifferential mappings. Pacific J. Math. **33**, 209–216 (1970)
18. Rockafellar, R.T.: On the maximality of sums of nonlinear monotone operators. Trans. Am. Math. Soc. **149**, 75–88 (1970)
19. Rockafellar, R.T.: Monotone operators and proximal point algorithm. SIAM J. Control. Optim. **14**, 877–898 (1976)
20. Takahashi, W.: Nonlinear Functional Analysis: Fixed Point Theory and Its Applications. Yokohama Publishers, Yokohama 2000
21. Takahashi, W.: Convex Analysis and Approximation of Fixed Points (in Japanese). Yokohama Publishers, Yokohama 2000

22. Takahashi, W.: Fixed point theorems and proximal point algorithms. In: Takahashi, W., Tanaka, T. (eds.) Nonlinear Analysis and Convex Analysis, pp. 471–481. Yokohama Publishers, Yokohama 2003
23. Takahashi, W.: Convergence theorems for nonlinear projections in Banach spaces (in Japanese). RIMS Kokyuroku, vol. 1396, pp. 49–59 (2004)

Adv. Math. Econ. 10, 65–89 (2007)

Advances in
MATHEMATICAL
ECONOMICS

Golden optimal policy in calculus of variation and dynamic programming

Seiichi Iwamoto

Department of Economic Engineering, Graduate School of Economics, Kyushu University, Fukuoka 812-8581, Japan
(e-mail: iwamoto@en.kyushu-u.ac.jp)

Received: April 22, 2006
Revised: September 12, 2006

JEL classification: C61, D81

Mathematics Subject Classification (2000): 90C39, 90C40, 90A43

Abstract. This paper discusses four dynamic optimization problems on an infinite continuous time interval from a viewpoint of Golden optimality. The problem is whether an optimal policy is Golden or not. We solve two control processes with quadratic cost criterion and two allocation processes with discounted square-root reward criterion. Both processes have a linear dynamics. It is shown that one cotrol process does not admit a Golden optimal policy. The other three processes have a Golden optimal policy. Further we illustrate the Golden optimal trajectories through three approaches: (i) one-parametric method, (ii) Euler equation and (iii) Bellman equation.

Key words: golden policy, optimal policy, Euler equation, Bellman equation, golden optimality, control process, allocation process, golden section

1. Introduction

E. Phelps says in "*Golden Rule of Accumulation: A Fable for Growthmen*" [12] as follows: The King commended the task force for its informative and stimulating report. He invited all his subjects to join in search of an optimal investment policy. Solovian theorists considered dozen of fiscal devices for their efficiency, equity and effectiveness. Mathematicians, leading the quest for a growth strategy, grappled with extremals, functionals and Hamiltonians. Yet nothing practical emerged. Then a policy-maker was heard to say, "Forget grand optimality. Solovians are a simple people. We need a simple policy..."

On the other hand, this paper discusses an optimality. We are concerned with Golden optimality [8–11] for both control processes and allocation processes on continuous time. Here we also need a simple policy. However, it is the Golden optimal policy. A trajectory is called Golden if any current state moves to a next state constantly repeating a *Golden section* [4, 13] in unit time. A policy is called Golden if it, together with a relevant dynamics, generates a Golden trajectory. We direct our attention to the Golden policy as well as optimal policy.

We consider four dynamic optimization problems on time interval $[0, \infty)$ through three approaches. Two control processes with a quadratic criterion and two allocation processes with a discounted square-root criterion are solved in § 3 and in § 4, respectively. Our three approaches are (i) one-variable optimization method, (ii) calculus of variation and (iii) dynamic programming method. While (i) and (ii) concern a Golden trajectory, (iii) concerns a Golden policy. We show that one control process does not admit any Golden optimal policy and that the other three processes have a Golden optimal policy.

2. Golden trajectories

A real number

$$\phi = \frac{1+\sqrt{5}}{2} \approx 1.618$$

is called *Golden number* [4, 5, 13]. It is the larger of the two solutions to quadratic equation

$$x^2 - x - 1 = 0. \tag{1}$$

Sometimes (1) is called *Fibonacci quadratic equation* [5]. The Fibonacci quadratic equation has two real solutions: ϕ and its *conjugate* $\overline{\phi} := 1 - \phi$. We note that

$$\phi + \overline{\phi} = 1, \quad \phi \cdot \overline{\phi} = -1.$$

Further we have

$$\phi^2 = 1 + \phi, \quad \overline{\phi}^2 = 2 - \phi$$
$$\phi^2 + \overline{\phi}^2 = 3, \quad \phi^2 \cdot \overline{\phi}^2 = 1.$$

Definition 21. *A differentiable function* $x : [0, \infty) \to R^1$ *is called Golden if and only if either*

$$\frac{\dot{x}(t)}{x(t)} = \log(\phi - 1) \quad or \quad \frac{\dot{x}(t)}{x(t)} = \log(2 - \phi). \tag{2}$$

Lemma 21. *A Golden function* x *is either*

$$x(t) = x(0)e^{-t\log\phi} \quad or \quad x(t) = x(0)e^{-t\log(1+\phi)}.$$

We remark that in Fig. 1

$$e^{-t\log\phi} = e^{t\log(\phi-1)} = \phi^{-t} = (\phi - 1)^t,$$

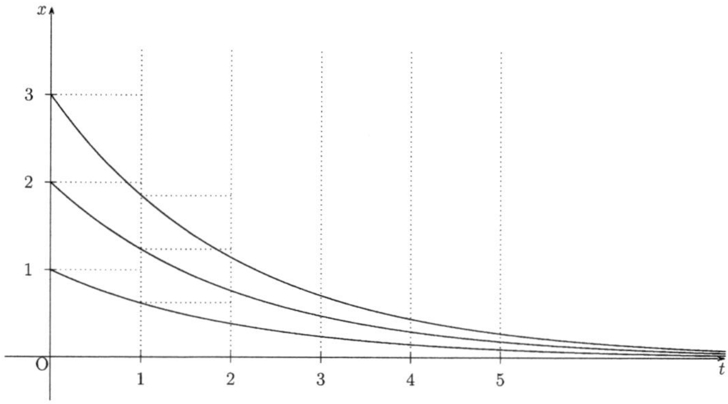

Fig. 1. Golden trajectories $x = c(\phi - 1)^t$, $c = 1, 2, 3$

where

$$\log \phi \approx 0.481, \quad \phi - 1 = \phi^{-1} \approx 0.618$$

and that in Fig. 2

$$e^{-t \log(1+\phi)} = e^{t \log(2-\phi)} = (1 + \phi)^{-t} = (2 - \phi)^t,$$

where

$$\log(1 + \phi) \approx 0.962, \quad 2 - \phi = (1 + \phi)^{-1} \approx 0.382.$$

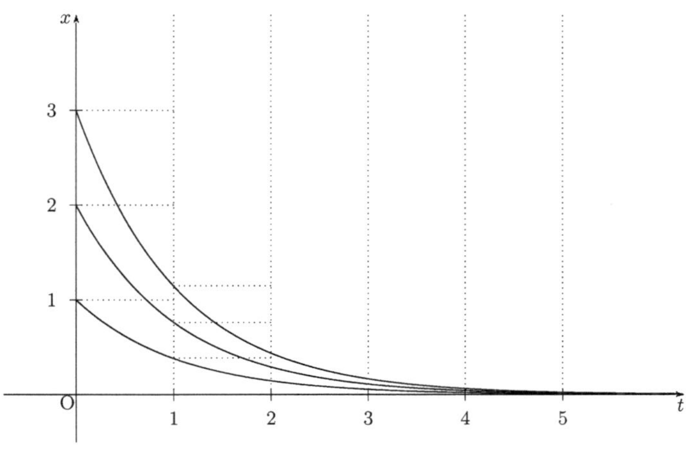

Fig. 2. Golden trajectories $x = c(2 - \phi)^t$, $c = 1, 2, 3$

Let us introduce a controlled linear dynamics with real parameter b as follows [2]:

$$\dot{x} = bx + u \quad \text{on } [0, \infty), \tag{3}$$

where $u : R^1 \to R^1$ is called *control function* or *control*. If $u(x) = px$ (resp. $px + q$), the control is called *proportional (resp. linear)*, where p, q are real constants. A differentiable solution x to (3) is called *trajectory*.

Definition 22. *A proportional control u on dynamics* (3) *is called Golden if and only if it generates a Golden trajectory.*

Lemma 22. *A proportional control $u = px$ on* (3) *is Golden if and only if*

$$p = -b + \log(\phi - 1) \quad \textit{or} \quad p = -b + \log(2 - \phi).$$

3. Control processes

This section minimizes two quadratic cost functions

$$\int_0^\infty \left(x^2 + u^2\right) dt \quad \text{and} \quad \int_0^\infty \left(\dot{x}^2 + u^2\right) dt$$

under a common additive dynamics

$$\dot{x} = bx + u$$

(see [2, 3]).

3.1. First quadratic criterion

Let C^1 be the set of all continuously differentiable functions on the nonnegative half-line $[0, \infty)$:

$$C^1 = \left\{x = x(t) \mid x : [0, \infty) \to R^1 \text{ continuously differentiable}\right\}.$$

First we take the quadratic criterion

$$I(x) = \int_0^\infty \left[x^2 + (bx - \dot{x})^2\right] dt.$$

Now we consider a variational problem for a given real constant c:

$$\text{MP}_1(\text{c}) \quad \text{minimize } I(x) \text{ subject to (i) } x \in C^1, \text{ (ii) } x(0) = c.$$

Let us evaluate a few special trajectories:

1. A constant $x(t) = c$ yields $I(x) = \infty$.
2. A proportional $x(t) = ce^{-at}$ $(0 < a < \infty)$ yields

$$\begin{aligned} I(x) &= c^2\left[1 + (a+b)^2\right]\int_0^\infty e^{-2at}dt \\ &= c^2\frac{1 + (a+b)^2}{2a}. \end{aligned}$$

3. Since

$$\min_{0<a<\infty} \frac{1 + (a+b)^2}{2a}$$

is attained at $\hat{a} = \sqrt{1+b^2}$, we have the minimum value $b + \sqrt{1+b^2}$. Thus, a proportional $\hat{x}(t) = ce^{-\sqrt{1+b^2}\,t}$ yields the optimal value

$$I(\hat{x}) = (b + \sqrt{1+b^2})c^2,$$

in the class of proportional trajectories.

Since $1 > \log(1+\phi) \approx 0.962$
neither

$$\sqrt{1+b^2} = \log(1+\phi)$$

nor

$$\sqrt{1+b^2} = \log\phi$$

has solution. This implies that the proportional trajectory $\hat{x}(t) = ce^{-\sqrt{1+b^2}\,t}$ should not be Golden. Thus, we have no Golden optimal trajectory in $\mathrm{MP}_1(\mathrm{c})$.

3.1.1. Euler equation

Let us consider an extremal problem

$$\mathrm{EP}_1(\mathrm{c}) \quad \text{extremize} \int_0^\infty f(t, x, \dot{x})dt \text{ subject to (i) } x \in C^1, \text{ (ii) } x(0) = c.$$

Then an extremal curve $x = x(\cdot)$ satisfies a variational equation, *Euler equation*,

$$\frac{d}{dt}f_{\dot{x}} - f_x = 0 \tag{4}$$

and a boundary condition at infinity ∞, *transvesality condition*,

$$\lim_{t\to\infty} f_{\dot{x}}(t, x(t), \dot{x}(t)) = 0. \tag{5}$$

Formally both equations are shown as follows (see [6]). Let $\eta = \eta(\cdot)$ be any C^1-function satisfying $\eta(0) = 0$. Then $y := x + \epsilon\eta$ is feasible for any $\epsilon \in R^1$. Let us define

$$J(\epsilon) := \int_0^\infty f(t, y, \dot{y})dt = \lim_{T\to\infty} \int_0^T f(t, y, \dot{y})dt.$$

Then $J(\cdot)$ must take a minimum value at $\epsilon = 0$ for any such η. This implies $J'(0) = 0$.

Let us now calculate $J'(0) = \lim_{h\to 0} \dfrac{J(h) - J(0)}{h}$. First we note that

$$\frac{J(h) - J(0)}{h} = \lim_{T\to\infty} \int_0^T g(t)dt,$$

where

$$g(t) := g(t; h) = \frac{f(t, x + h\eta, \dot{x} + h\dot{\eta}) - f(t, x, \dot{x})}{h}.$$

From the mean value theorem, there exists θ $(0 \leq \theta \leq 1)$ satisfying

$$g(t) = f_x\eta + f_{\dot{x}}\dot{\eta},$$

where

$$\begin{aligned} f_x &= f_x(t, x + \theta h\eta, \dot{x} + \theta h\dot{\eta}) \\ f_{\dot{x}} &= f_x(t, x + \theta h\eta, \dot{x} + \theta h\dot{\eta}). \end{aligned}$$

Hence, from integration by parts and $\eta(0) = 0$, it holds that

$$\begin{aligned} \int_0^T g(t)dt &= \int_0^T (f_x\eta + f_{\dot{x}}\dot{\eta})dt \\ &= \int_0^T \left(f_x\eta - \frac{d}{dt}f_{\dot{x}}\eta\right)dt + [f_{\dot{x}}\eta(t)]_0^T \\ &= \int_0^T \left(f_x - \frac{d}{dt}f_{\dot{x}}\right)\eta dt + f_{\dot{x}}\eta(T). \end{aligned}$$

Then we have

$$\lim_{T\to\infty} \int_0^T g(t)dt = \lim_{T\to\infty} \int_0^T \left(f_x - \frac{d}{dt}f_{\dot{x}}\right)\eta dt + \lim_{T\to\infty} f_{\dot{x}}\eta(T).$$

Thus

$$\begin{aligned} J'(0) &= \lim_{h\to 0} \frac{J(h)-J(0)}{h} \\ &= \lim_{h\to 0}\lim_{T\to\infty}\int_0^T g(t)dt \\ &= \lim_{h\to 0}\lim_{T\to\infty}\int_0^T \left(f_x - \frac{d}{dt}f_{\dot{x}}\right)\eta dt + \lim_{h\to 0}\lim_{T\to\infty} f_{\dot{x}}\eta(T) \\ &= \lim_{T\to\infty}\int_0^T \lim_{h\to 0}\left(f_x - \frac{d}{dt}f_{\dot{x}}\right)\eta dt + \lim_{T\to\infty}\lim_{h\to 0} f_{\dot{x}}\eta(T). \end{aligned}$$

Consequently, it holds that

$$J'(0) = \lim_{T\to\infty}\int_0^T \left(f_x - \frac{d}{dt}f_{\dot{x}}\right)\eta dt + \lim_{T\to\infty} f_{\dot{x}}\eta(T),$$

where

$$\begin{aligned} f_x &= f_x(t,x,\dot{x}) \\ f_{\dot{x}} &= f_{\dot{x}}(t,x,\dot{x}). \end{aligned}$$

Since $J'(0)$ must vanish for any C^1-function η satisfying $\eta(0)=0$, we have

$$\begin{aligned} f_x - \frac{d}{dt}f_{\dot{x}} &= 0 \quad \text{on } [0,\infty) \\ \lim_{t\to\infty} f_{\dot{x}}(t,x(t),\dot{x}(t)) &= 0. \end{aligned}$$

Thus, both Euler equation and the transversality condition are derived. □

Now let us take the quadratic criterion

$$f(t,x,\dot{x}) = x^2 + (bx-\dot{x})^2.$$

Then the Euler equation (4) implies

$$\ddot{x} - a^2x = 0, \qquad (6)$$

where

$$a = \sqrt{1+b^2}.$$

The transvesality condition (5) implies

$$\lim_{t\to\infty}[\dot{x}(t) - bx(t)] = 0. \qquad (7)$$

By solving the system of three equations, initial condition (ii), (6), and (7), we have

$$x(t) = ce^{-at}.$$

3.1.2. Bellman equation

Let us now consider a control process with constant parameter b $(-\infty < b < \infty)$:

$$\mathrm{C}(c) \qquad \begin{array}{ll} \text{minimize} & \displaystyle\int_0^\infty \left(x^2 + u^2\right) dt \\ \text{subject to} & \text{(i) } \dot{x} = bx + u \qquad\qquad 0 \le t < \infty \\ & \text{(ii) } u \in C^1 \\ & \text{(iii) } x(0) = c. \end{array}$$

Let $v(c)$ be the minimum value. When $c = 0$, the feasible $x(t) = u(t) = 0$ yields null integral value. Hence $v(0) = 0$. The value function $v = v(x)$ satisfies the Bellman equation [1–3, 7] :

$$\min_{-\infty<u<\infty} \left[x^2 + u^2 + v'(x)(bx + u)\right] = 0, \quad v(0) = 0. \tag{8}$$

Since $\dfrac{d}{du}[\cdots] = 0$ yields $u = -\dfrac{v'(x)}{2}$, we have

$$\min_{-\infty<u<\infty} [\cdots] \ = \ x^2 + bx \cdot v'(x) - \frac{1}{4}\left(v'(x)\right)^2.$$

Thus equation (8) implies

$$\frac{1}{4}\left(v'(x)\right)^2 - bx \cdot v'(x) - x^2 = 0.$$

Assuming a quadratic form $v(x) = kx^2$ $(k > 0)$, the coefficient k must satisfy a quadratic equation

$$k^2 - 2bk - 1 = 0.$$

Thus, we have

$$k = b + \sqrt{b^2 + 1}\,.$$

Hence, we have the following result.

Lemma 31. *A pair of proportional control and quadratic value function*

$$\hat{u}(x) = -kx, \quad v(x) = kx^2$$

satisfies the Bellman equation

$$\min_{-\infty<u<\infty} \left[x^2 + u^2 + v'(x)(bx + u)\right] = 0.$$

Then the dynamical system

$$\dot{x} = -\sqrt{b^2+1}\,x, \quad x(0) = c,$$

yields an trajectory

$$\hat{x}(t) = ce^{-\sqrt{b^2+1}\,t}.$$

The proportional policy $\hat{u}$ generates an optimal trajectory $\hat{x}$, which is not Golden. Thus, the control process $\mathrm{C}(c)$ does not admit a Golden optimal policy. However, when $b = 1/2$, the coefficient k reduces to the Golden ratio $k = \phi$.

3.2. Second quadratic criterion

Second we take the quadratic criterion

$$J(x) \;=\; \int_0^\infty \left[\dot{x}^2 + (bx - \dot{x})^2\right] dt \quad \text{on } C^1$$

where $0 \le b < \infty$.

Now we consider a mathematical programming problem for a given real constant c:

MP$_2$(c) minimize $J(x)$ subject to (i) $x \in C^1$, (ii) $x(0) = c$.

Let us evaluate a few special trajectories:

1. A constant $x(t) = c$ yields $J(x) = \infty$.
2. A proportional $x(t) = ce^{-at}$ yields

$$\begin{aligned} I(x) &= c^2\left[a^2 + (a+b)^2\right] \int_0^\infty e^{-2at}\,dt \\ &= c^2 \frac{a^2 + (a+b)^2}{2a} \quad (0 < a < \infty). \end{aligned}$$

3. Since

$$\min_{0<a<\infty} \frac{a^2 + (a+b)^2}{2a}$$

is attained at $\hat{a} = \dfrac{b}{\sqrt{2}}$, we have the minimum value $(1+\sqrt{2})b$.

Thus, a proportional $\hat{x}(t) = ce^{-(b/\sqrt{2})t}$ yields the optimal value

$$I(\hat{x}) \;=\; (1+\sqrt{2})bc^2$$

in the class of proportional trajectories.

The proportional trajectory $\hat{x}(t) = ce^{-(b/\sqrt{2})t}$ is Golden if and only if either of the following two cases holds:

- Case (i) $\dfrac{b}{\sqrt{2}} = \log(1+\phi)$
  implies
  $$b_1 = \sqrt{2}\log(1+\phi) \approx 1.361$$
  Then we have a Golden trajectory
  $$\hat{x}_1(t) = ce^{-t\log(1+\phi)}.$$
- Case (ii) $\dfrac{b}{\sqrt{2}} = \log\phi$
  implies
  $$b_2 = \sqrt{2}\log\phi \approx 0.681.$$
  Then we have a Golden trajectory
  $$\hat{x}_2(t) = ce^{-t\log\phi}.$$

Thus, we have the two Golden optimal trajectories $\hat{x}_1$ and $\hat{x}_2$.

3.2.1. Euler equation

We consider an extremal problem

$$\mathrm{EP}_2(c) \quad \text{extremize} \int_0^\infty f(x,\dot{x})dt \text{ subject to (i) } x \in C^1, \text{ (ii) } x(0) = c$$

where

$$f(x,\dot{x}) = \dot{x}^2 + (bx - \dot{x})^2.$$

Then the Euler equation (4) implies

$$\ddot{x} - a^2 x = 0, \tag{9}$$

where

$$a = \frac{b}{\sqrt{2}}.$$

The transvesality condition (5) implies

$$\lim_{t\to\infty}(2\dot{x}(t) - bx(t)) = 0. \tag{10}$$

By solving the system of three equations, initial condition (ii), (9), and (10), we have

$$x(t) = ce^{-at}.$$

3.2.2. Bellman equation

Let us now consider a control process with constant parameter b ($-\infty < b < \infty$):

$$
\mathrm{C}'(c) \qquad \begin{array}{ll} \text{minimize} & \displaystyle\int_0^\infty \left(\dot{x}^2 + u^2\right) dt \\ \text{subject to} & \text{(i) } \dot{x} = bx + u \qquad\qquad 0 \le t < \infty \\ & \text{(ii) } u \in C^1 \\ & \text{(iii) } x(0) = c. \end{array}
$$

Let $v(c)$ be the minimum value. Then the value function $v = v(x)$ satisfies the Bellman equation with an initial condition:

$$
\min_{-\infty<u<\infty} \left[(bx+u)^2 + u^2 + v'(x)(bx+u)\right] = 0, \quad v(0) = 0. \qquad (11)
$$

Since $\dfrac{d}{du}[\cdots] = 0$ yields $u = -\dfrac{1}{4}\left(2bx + v'(x)\right)$, we have

$$
\min_{-\infty<u<\infty} [\cdots] = \frac{1}{2}\left[(bx)^2 + bxv'(x) - \frac{1}{4}(v'(x))^2\right].
$$

Thus, equation (11) implies

$$
\frac{1}{4}\left(v'(x)\right)^2 - bxv'(x) - b^2x^2 = 0.
$$

Assuming a quadratic form $v(x) = kx^2$ ($k > 0$), the coefficient k satisfies a quadratic equation

$$
k^2 - 2bk - b^2 = 0.
$$

Hence, it follows that

$$
k = (1+\sqrt{2})b.
$$

We have

Lemma 32. *A pair of proportional control and quadratic value function*

$$
\hat{u}(x) = px, \quad v(x) = kx^2
$$

where

$$
p = -\frac{1}{2}(2+\sqrt{2})b, \quad k = (1+\sqrt{2})b
$$

satisfies the Bellman equation

$$
\min_{-\infty<u<\infty} \left[(bx+u)^2 + u^2 + v'(x)(bx+u)\right] = 0.
$$

Then the dynamical system

$$\dot{x} = -\frac{b}{\sqrt{2}}x, \quad x(0) = c,$$

yields an optimal trajectory

$$\hat{x}(t) = ce^{-(b/\sqrt{2})t}.$$

- Case (i) When $b = \sqrt{2}\log(1+\phi) \approx 1.361$ the optimal proportional policy $\hat{u}(x) = -(1+\sqrt{2})\log(1+\phi)\,x \approx -2.323\,x$ is Golden. It generates the Golden trajectory

$$\hat{x}(t) = ce^{-t\log(1+\phi)}.$$

- Case (ii) When $b = \sqrt{2}\log\phi \approx 0.681$ the optimal proportional policy $\hat{u}(x) = -(1+\sqrt{2})\log\phi\,x \approx -1.162\,x$ is Golden. It generates the Golden trajectory

$$\hat{x}(t) = ce^{-t\log\phi}.$$

4. Allocation processes

This section maximizes two discounted square-root reward functions

$$\int_0^\infty e^{-\rho t}(\sqrt{x}+\sqrt{u})dt \quad \text{and} \quad \int_0^\infty e^{-\rho t}\sqrt{u}\,dt$$

under a common subtractive dynamics

$$\dot{x} = bx - u.$$

4.1. First square-root criterion

Let two nonnegative constants ρ and b be given in $\rho - b \geq \sqrt{b}$. Then we have an inequality with equality condition as follows.

Lemma 41. *It holds that*

$$\frac{1+\sqrt{b+x}}{\rho+\frac{1}{2}x} \leq \frac{1}{\sqrt{1+2\rho-b}-1} \quad \text{on } [-b,\infty).$$

The sign of equality holds if and only if $x = 2(1+\rho-b-\sqrt{1+2\rho-b})$ *(see Fig. 3).*

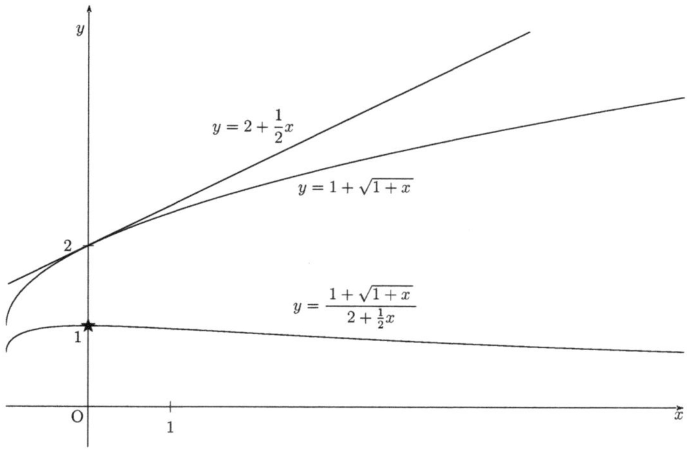

Fig. 3. Allocation ratio $y = \dfrac{1+\sqrt{1+x}}{2+\frac{1}{2}x}$, ★ denotes maximum point

Let C' be the set of all continuously differentiable functions satisfying $\dot{x}(t) \leq bx(t),\ x(t) \geq 0$ on the nonnegative half-line:

$$C' = \{x = x(t) \mid x : [0, \infty) \to R^1 \text{ continuously differentiable},\ \dot{x} \leq bx,\ \ x \geq 0\}.$$

First we take the first square-root criterion

$$K(x) = \int_0^\infty e^{-\rho t}(\sqrt{x} + \sqrt{bx - \dot{x}})dt \quad \text{on } C'.$$

Now we consider a variational problem for a given real constant c:

$$\text{MP}_3(\text{c}) \quad \text{Maximize } K(x) \text{ subject to (i) } x \in C',\ \text{(ii) } x(0) = c.$$

Let us evaluate a few special trajectories:

1. A constant $x(t) = c$ yields $K(x) = (1+\sqrt{b})\dfrac{\sqrt{c}}{\rho}$.
2. A proportional $x(t) = ce^{-at}$ $(-b \leq a < \infty)$ yields

$$\begin{aligned} K(x) &= \sqrt{c}(1+\sqrt{b+a})\int_0^\infty e^{-(\rho+\frac{1}{2}a)t}dt \\ &= \sqrt{c}\,\frac{1+\sqrt{b+a}}{\rho+\frac{1}{2}a}. \end{aligned}$$

3. Since

$$\underset{-b\leq a<\infty}{\text{Max}} \frac{1+\sqrt{b+a}}{\rho+\frac{1}{2}a}$$

is attained at $a^* = 2(1+\rho-b-\sqrt{1+2\rho-b})$, we have the maximum value

$$\frac{1}{\sqrt{1+2\rho-b}-1}.$$

Thus a proportional $x^*(t) = ce^{-a^*t}$ yields the optimal value

$$K(x^*) = \frac{\sqrt{c}}{\sqrt{1+2\rho-b}-1},$$

in the class of proportional trajectories.

4.1.1. Case $b = 1$

Let us consider the case $b = 1$. Then, under the condition $\rho \geq 2$, we have the proportional trajectory $x^*(t) = ce^{-2(\rho-\sqrt{2\rho})t}$. It is Golden if and only if either of the following two cases holds.

- Case (i) $2(\rho-\sqrt{2\rho}) = \log(1+\phi)$ implies

$$\rho_1 = \frac{1}{2}(1+\sqrt{1+\log(1+\phi)})^2 \approx 2.882$$

  Then we have a Golden trajectory $\hat{x}_1(t) = ce^{-t\log(1+\phi)}$.
- Case (ii) $2(\rho-\sqrt{2\rho}) = \log\phi$ implies

$$\rho_2 = \frac{1}{2}(1+\sqrt{1+\log\phi})^2 \approx 2.458$$

  Then we have a Golden trajectory $\hat{x}_2(t) = ce^{-t\log\phi}$.

Thus, we have the two Golden optimal trajectories $\hat{x}_1$ and $\hat{x}_2$.

4.1.2. Case $b = 0$

Let us consider the case $b = 0$. Since $2(1+\rho-\sqrt{1+2\rho}) = (\sqrt{1+2\rho}-1)^2$, we have the proportional trajectory $x^*(t) = ce^{-(\sqrt{1+2\rho}-1)^2t}$ under the condition $\rho \geq 0$. It is Golden if and only if either of the following two cases holds.

- Case (i) $(\sqrt{1+2\rho}-1)^2 = \log(1+\phi)$
  implies

$$\rho_1 = \sqrt{\log(1+\phi)} + \frac{1}{2}\log(1+\phi) \approx 1.462$$

  Then we have a Golden trajectory $\hat{x}_1(t) = ce^{-t\log(1+\phi)}$.
- Case (ii) $(\sqrt{1+2\rho}-1)^2 = \log\phi$
  implies

$$\rho_2 = \sqrt{\log\phi} + \frac{1}{2}\log\phi \approx 0.934$$

  Then we have a Golden trajectory $\hat{x}_2(t) = ce^{-t\log\phi}$.

Thus, we have the two Golden optimal trajectories $\hat{x}_1$ and $\hat{x}_2$.

4.2. Euler equation

Let us consider an extremal problem

$$\text{EP}_3(c) \quad \text{extremize} \int_0^\infty f(t,x,\dot{x})dt \text{ subject to (i) } x \in C', \text{ (ii) } x(0) = c.$$

Let us take the square-root criterion

$$f(t,x,\dot{x}) = e^{-\rho t}(\sqrt{x} + \sqrt{bx-\dot{x}}).$$

Then we have

$$\begin{aligned} f_x &= \frac{1}{2}e^{-\rho t}\left(\frac{1}{\sqrt{x}} + \frac{b}{\sqrt{x-\dot{x}}}\right) \\ f_{\dot{x}} &= -\frac{1}{2}e^{-\rho t}\frac{1}{\sqrt{bx-\dot{x}}} \\ \frac{d}{dt}f_{\dot{x}} &= \frac{1}{2}e^{-\rho t}\left(\frac{\rho}{\sqrt{bx-\dot{x}}} + \frac{b\dot{x}-\ddot{x}}{2(bx-\dot{x})^{3/2}}\right). \end{aligned}$$

The Euler equation (4) implies

$$\frac{\rho-b}{\sqrt{bx-\dot{x}}} + \frac{b\dot{x}-\ddot{x}}{2(x-\dot{x})^{3/2}} = \frac{1}{\sqrt{x}}. \tag{12}$$

The transversality condition (5) implies

$$\lim_{t\to\infty}\frac{1}{\sqrt{bx(t)-\dot{x}(t)}} = 0. \tag{13}$$

Let us now find a solution of a form $x(t) = ce^{-at}$. Substituting $\dot{x} = -cae^{-at}$, $\ddot{x} = ca^2e^{-at}$ into (12), the coefficient a must satisfy

$$\frac{\rho - b}{\sqrt{b+a}} - \frac{ba + a^2}{2(b+a)^{3/2}} = 1,$$

namely,

$$\rho - b - \frac{a}{2} = \sqrt{b+a}. \tag{14}$$

Solving (14) yields

$$a = 2(1 + \rho - b - \sqrt{1 + 2\rho - b}).$$

Thus, we have a solution to (12), (13):

$$x(t) = ce^{-at}.$$

4.3. Bellman equation

Let us now consider a control process with constant parameter b $(-\infty < b < \infty)$:

$$\mathrm{A}(c) \qquad \begin{array}{lll} \text{Maximize} & \displaystyle\int_0^\infty e^{-\rho t} f(x,u)dt & \\ \text{subject to} & \text{(i) } \dot{x} = g(x,u) & \\ & \text{(ii) } x \in C^1, \ u \in U(x) & 0 \le t < \infty \\ & \text{(iii) } x(0) = c. & \end{array}$$

Let $v(c)$ be the maximum value. Then the value function $v = v(x)$ satisfies Bellman equation:

$$\rho v(x) = \underset{u \in U(x)}{\mathrm{Max}} \left[f(x,u) + v'(x)g(x,u) \right]. \tag{15}$$

Formally this is derived as follows (see [1–3]). Let us take any small $\Delta > 0$. We define for any feasible paired (x, u)-process a new paired (y, w)-process as follows:

$$y(t) := x(t + \Delta), \quad w(t) := u(t + \Delta), \ \ t \in [0, \infty).$$

Then the process $y = \{y(\cdot)\}_{[0,\infty)}$ satisfies

$$\begin{array}{ll} \text{(i)}' \ \dot{y} = h(y,w) & \\ \text{(ii)}' \ y \in C^1, \ w \in U(y) & 0 \le t < \infty \\ \text{(iii)}' \ y(0) = x(\Delta). & \end{array}$$

Conversely, concatenating the paired process $(x(\cdot), u(\cdot))$ on time interval $[0, \Delta]$ for any paired process$(y(\cdot), w(\cdot))$ satisfying (i)$'$, (ii)$'$, (iii)$'$, we can construct a paired (x, u)-process on the total time interval $[0, \infty)$ satisfying conditions (i), (ii), (iii). Thus, there is a one-to-one correspondence between (x, u)-process and (y, w)-process.

First we note that

$$\int_0^\infty e^{-\rho t} f(x, u)dt \;=\; \int_0^\Delta e^{-\rho t} f(x, u)dt + e^{-\rho\Delta} \int_0^\infty e^{-\rho t} f(y, w)dt.$$

From the mean value theorem, there exists θ $(0 \le \theta \le 1)$ satisfying

$$\int_0^\Delta h(t)dt = h(\theta\Delta)\Delta,$$

where

$$h(t) = e^{-\rho t} f(x, u).$$

Thus, we have

$$\begin{aligned} v(c) \;&\ge\; \int_0^\infty e^{-\rho t} f(x, u)dt \\ &=\; h(\theta\Delta)\Delta + e^{-\rho\Delta} \int_0^\infty e^{-\rho t} f(y, w)dt. \end{aligned}$$

From the preceding construction of the paired (x, u)-process on $[0, \infty)$ for any paired process$(y(\cdot), w(\cdot))$ satisfying (i)$'$, (ii)$'$, (iii)$'$, we have

$$v(c) \;\ge\; h(\theta\Delta)\Delta + e^{-\rho\Delta} v(x(\Delta)).$$

Applying the mean value theorem two times, there exist η, ξ $(0 \le \eta, \xi \le 1)$ satisfying

$$\begin{aligned} &h(\theta\Delta)\Delta + e^{-\rho\Delta} v(x(\Delta)) \\ &= h(\theta\Delta)\Delta + \left[1 - \rho\Delta + \frac{1}{2!} e^{-\xi\rho\Delta} (\rho\Delta)^2\right] \\ &\quad \times \left[v(c) + v'(x(\eta\Delta))\dot{x}(\eta\Delta)\Delta\right]. \end{aligned} \tag{16}$$

Thus, we have

$$v(c) \;\ge\; h(\theta\Delta)\Delta + \left[1 - \rho\Delta + \frac{1}{2!} e^{-\xi\rho\Delta} (\rho\Delta)^2\right]\left[v(c) + v'(x(\eta\Delta))\dot{x}(\eta\Delta)\Delta\right].$$

Subtracting $v(c)$, dividing Δ and taking a limit as $\Delta \downarrow 0$, we have for any feasible u

$$\rho v(c) \;\ge\; f(c, u) + v'(c) g(x, u).$$

Thus, we have

$$\rho v(c) \geq \underset{u \in U(c)}{\text{Max}} \left[f(c,u) + v'(c)g(c,u) \right].$$

Conversely, let us assume that a feasible (hence, optimal) paired (x, u)-process attains the maximum value function v. Then we have from (16) once again

$$\begin{aligned} v(c) &= \int_0^\infty e^{-\rho t} f(x,u)dt \\ &= h(\theta\Delta)\Delta + e^{-\rho\Delta} v(x(\Delta)) \\ &= h(\theta\Delta)\Delta + \left[1 - \rho\Delta + \frac{1}{2!} e^{-\xi\rho\Delta} (\rho\Delta)^2 \right] \left[v(c) + v'(x(\eta\Delta))\dot{x}(\eta\Delta)\Delta \right]. \end{aligned}$$

This in turn yields

$$\rho v(c) = f(c,u) + v'(c)g(x,u).$$

Thus, we have the converse inequality

$$\rho v(c) \leq \underset{u \in U(c)}{\text{Max}} \left[f(c,u) + v'(c)g(c,u) \right].$$

Thus, the Bellman equation (15) is valid. □

Let us now take a square-root criterion and a subtractive dynamics as follows:

$$f(x,u) = \sqrt{x} + \sqrt{u}, \quad g(x,u) = bx - u.$$

We solve an allocation process with a discounted total square-root criterion:

$$\mathrm{A}(c) \quad \begin{array}{lll} \text{Maximize} & \displaystyle\int_0^\infty e^{-\rho t}(\sqrt{x} + \sqrt{u})dt & \\ \text{subject to} & \text{(i) } \dot{x} = bx - u & \\ & \text{(ii) } x \in C^1, \ u \geq 0 & 0 \leq t < \infty \\ & \text{(iii) } x(0) = c. & \end{array}$$

Let us solve the Bellman equation:

$$\rho v(x) = \underset{u \geq 0}{\text{Max}} \left[\sqrt{x} + \sqrt{u} + v'(x)(bx - u) \right]. \tag{17}$$

Since $\frac{d}{du}[\cdots] = 0$ yields

$$\frac{1}{2\sqrt{u}} = v' \quad \text{i.e., } u = \frac{1}{4v'^2} \tag{18}$$

we have

$$\operatorname*{Max}_{u \geq 0} [\cdots] \;=\; \sqrt{x} + bxv' + \frac{1}{4v'}.$$

Thus, we have

$$\rho v(x) = \sqrt{x} + bxv'(x) + \frac{1}{4v'(x)}.$$

Let us find a square-root solution $v(x) = k\sqrt{x}$ $(k > 0)$ to (17). Then the coefficient k must satisfy

$$\left(\rho - \frac{b}{2}\right) k = 1 + \frac{1}{2k}. \tag{19}$$

Equation (19) has a solution

$$k = \frac{1}{\sqrt{1 + 2\rho - b} - 1}.$$

This together with (18) and (iii) yields

$$\dot{x} = -ax, \quad x(0) = c,$$

where

$$a = 2(1 + \rho - b - \sqrt{1 + 2\rho - b}).$$

Thus, we have the optimal trajectory

$$x^*(t) = ce^{-at}$$

and an optimal control

$$u^*(x) = (a + b)x.$$

4.3.1. Case $b = 1$

Let us consider the case $b = 1$. Then, under the condition $\rho \geq 2$, we have the proportional policy $u^*(x) = (1 + 2\rho - 2\sqrt{2\rho})\,x$. It generates an optimal trajectory $x^*(t) = ce^{-at}$, where $a = 2(\rho - \sqrt{2\rho})$.

- Case (i) When $\rho = \frac{1}{2}(1 + \sqrt{1 + \log(1 + \phi)})^2 \approx 2.882$, the optimal trajectory x^* is Golden $x_1^*(t) = ce^{-t\log(1+\phi)}$. This is generated by a Golden $u_1^*(x) = (1 + \log(1 + \phi))\,x \approx 1.962\,x$. Thus, we have the Golden optimal policy u_1^*.

- Case (ii) When $\rho = \frac{1}{2}(1 + \sqrt{1 + \log\phi})^2 \approx 2.458$, the optimal trajectory x^* is Golden $x_2^*(t) = ce^{-t\log\phi}$. This is generated by a Golden $u_2^*(x) = (1 + \log\phi)x \approx 1.481\,x$. Thus u_2^* is Golden optimal.

4.3.2. Case $b = 0$

Let us consider the case $b = 0$. Then, under the condition $\rho \geq 0$, we have the proportional policy $u^*(x) = ax$. It generates an optimal trajectory $x^*(t) = ce^{-at}$, where $a = (\sqrt{1 + 2\rho} - 1)^2$.

- Case (i) When $\rho = \sqrt{\log(1 + \phi)} + \frac{1}{2}\log(1 + \phi) \approx 1.462$, the optimal trajectory x^* is Golden $x_1^*(t) = ce^{-t\log(1+\phi)}$. This is generated by a Golden $u_1^*(x) = x\log(1+\phi) \approx 0.962\,x$. Thus, we have the Golden optimal policy u_1^*.
- Case (ii) When $\rho = \sqrt{\log\phi} + \frac{1}{2}\log\phi \approx 0.934$, the optimal trajectory x^* is Golden $x_2^*(t) = ce^{-t\log\phi}$. This is generated by a Golden $u_2^*(x) = x\log\phi \approx 0.481\,x$. Thus u_2^* is Golden optimal.

4.4. Second square-root criterion

Let ρ and b be given in $0 \leq b \leq \rho < \infty$. Then we have an inequality with equality condition as follows.

Lemma 42. *It holds that*

$$\frac{\sqrt{b + x}}{\rho + \frac{1}{2}x} \leq \frac{1}{\sqrt{2\rho - b}} \quad \textit{on } [-b, \infty).$$

The sign of equality holds if and only if $x = 2(\rho - b)$ (see Fig. 4).

Let us take the second square-root criterion

$$L(x) = \int_0^\infty e^{-\rho t}\sqrt{bx - \dot{x}}dt \quad \text{on } C'.$$

Now we consider a variational problem for a given real constant c:

$$\text{MP}_4(\text{c}) \quad \text{Maximize } L(x) \text{ subject to (i) } x \in C', \text{ (ii) } x(0) = c.$$

Let us evaluate a few special trajectories:

1. A constant $x(t) = c$ yields $L(x) = \dfrac{\sqrt{bc}}{\rho}$.

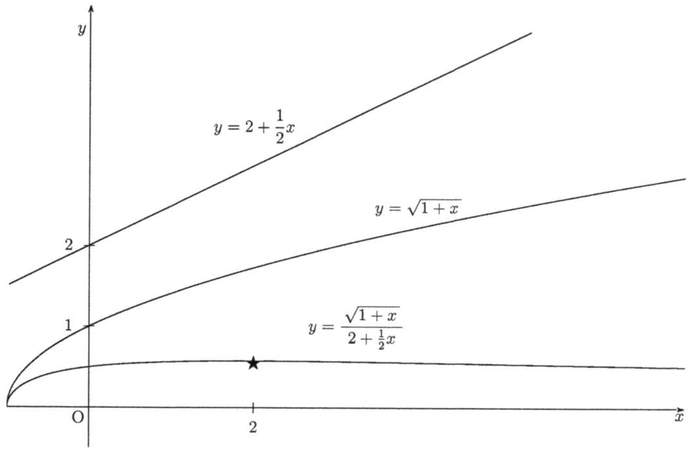

Fig. 4. Allocation ratio $y = \dfrac{\sqrt{1+x}}{2+\frac{1}{2}x}$, ★ denotes maximum point

2. A proportional $x(t) = ce^{-at}(-b \le a < \infty)$ yields

$$L(x) = \sqrt{c}\sqrt{b+a}\int_0^\infty e^{-(\rho+\frac{1}{2}a)t}dt = \sqrt{c}\frac{\sqrt{b+a}}{\rho+\frac{1}{2}a}.$$

3. Since

$$\underset{-b \le a < \infty}{\text{Max}} \frac{\sqrt{b+a}}{\rho+\frac{1}{2}a}$$

is attained at $a^* = 2(\rho - b)$, we have the maximum value

$$\frac{1}{\sqrt{2\rho - b}}.$$

Thus a proportional $x^*(t) = ce^{-2(\rho-b)t}$ yields the optimal value

$$L(x^*) = \frac{\sqrt{c}}{\sqrt{2\rho - b}},$$

in the class of proportional trajectories.

The proportional trajectory $x^*(t) = ce^{-2(\rho-b)t}$ is Golden if and only if either of the following two cases holds:

- Case (i) $2(\rho - b) = \log(1+\phi)$.
  implies

$$\rho_1 = b + \log\phi$$

  Then we have a Golden trajectory $\hat{x}_1(t) = ce^{-t\log(1+\phi)}$.
- Case (ii) $2(\rho - b) = \log\phi$
  implies

$$\rho_2 = b + \frac{1}{2}\log\phi.$$

  Then we have a Golden trajectory $\hat{x}_2(t) = ce^{-t\log\phi}$.

Thus, we have the two Golden optimal trajectories $\hat{x}_1$ and $\hat{x}_2$.

4.4.1. Euler equation

Now we consider an extremal problem

$$\mathrm{EP}_4(c) \quad \text{extremize} \int_0^{\infty} e^{-\rho t}\sqrt{bx - \dot{x}}dt \text{ subject to (i) } x \in C', \text{ (ii) } x(0) = c.$$

Then the second square-root criterion

$$f(t, x, \dot{x}) = e^{-\rho t}\sqrt{bx - \dot{x}}$$

yields

$$\begin{aligned} f_x &= \frac{1}{2}e^{-\rho t}\frac{b}{\sqrt{bx - \dot{x}}} \\ f_{\dot{x}} &= -\frac{1}{2}e^{-\rho t}\frac{1}{\sqrt{bx - \dot{x}}} \\ \frac{d}{dt}f_{\dot{x}} &= \frac{1}{2}e^{-\rho t}\left(\frac{\rho}{\sqrt{bx - \dot{x}}} + \frac{b\dot{x} - \ddot{x}}{2(bx - \dot{x})^{3/2}}\right). \end{aligned}$$

The Euler equation (4) implies

$$\rho + \frac{b\dot{x} - \ddot{x}}{2(bx - \dot{x})} = b,$$

namely,

$$\ddot{x} - (3b - 2\rho)\dot{x} + 2b(b - \rho) = 0. \tag{20}$$

The transversality condition (5) implies

$$\lim_{t\to\infty} \frac{e^{-\rho t}}{\sqrt{bx - \dot{x}}} = 0. \tag{21}$$

Let us now find a solution to (20), (ii), (21). Equation (20) has a general solution

$$x(t) = Ae^{-2(\rho-b)t} + Be^{bt},$$

where A, B are real constants. Thus, we have a desired solution

$$x(t) = ce^{-2(\rho-b)t}.$$

4.4.2. Bellman equation

Let us take a square-root criterion

$$f(x, u) = \sqrt{u}$$

We solve an allocation process with a discouted total square-root criterion:

$$\mathrm{A}'(c) \quad \begin{array}{ll} \text{Maximize} & \displaystyle\int_0^\infty e^{-\rho t}\sqrt{u}dt \\ \text{subject to} & \text{(i) } \dot{x} = bx - u \qquad\qquad 0 \le t < \infty \\ & \text{(ii) } x \in C^1, \ u \ge 0 \\ & \text{(iii) } x(0) = c. \end{array}$$

Let us solve the Bellman equation:

$$\rho v(x) = \underset{u \ge 0}{\mathrm{Max}} \left[\sqrt{u} + v'(x)(bx - u)\right]. \tag{22}$$

Since $\dfrac{d}{du}[\cdots] = 0$ yields

$$\frac{1}{2\sqrt{u}} = v', \quad \text{i.e., } u = \frac{1}{4v'^2} \tag{23}$$

we have

$$\underset{u \ge 0}{\mathrm{Max}}\,[\cdots] \ = \ bxv'(x) + \frac{1}{4v'}.$$

Thus, equation (22) reduces to

$$\rho v(x) = bxv'(x) + \frac{1}{4v'(x)}. \tag{24}$$

We will find a square-root solution $v(x) = k\sqrt{x}$ $(k > 0)$ to (24). Thus, the coefficient k must satisfy

$$\rho k = \frac{bk}{2} + \frac{1}{2k}. \tag{25}$$

Equation (25) has a solution

$$k = \frac{1}{\sqrt{2\rho - b}}.$$

This together with (23) and (iii) yields

$$\dot{x} = -ax, \quad x(0) = c$$

where $a = 2(\rho - b)$. Thus, we have the optimal trajectory

$$x^*(t) = ce^{-at}$$

and an optimal control

$$u^*(x) = (b + a)x.$$

4.4.3. Two cases

Let us specify two cases where the allocation process $A'(c)$ has a Golden optimal policy for any given $b \geq 0$.

- Case (i) When $\rho = b + \log\phi$, the optimal trajectory x^* is Golden $x_1^*(t) = ce^{-t\log(1+\phi)}$. This is generated by a Golden $u_1^*(x) = (b + 2\log\phi)x$. Thus we have the Golden optimal policy u_1^*.
- Case (ii) When $\rho = b + \frac{1}{2}\log\phi$, the optimal trajectory x^* is Golden $x_2^*(t) = ce^{-t\log\phi}$. This is generated by a Golden $u_2^*(x) = (b + \log\phi)\,x$. Thus u_2^* is Golden optimal.

References

1. Bellman, R.E.: Dynamic Programming. Princeton University Press, NJ 1957
2. Bellman, R.E.: Introduction of the Mathematical Theory of Control Processes, vol. 1, Linear Equations and Quadratic Criteria; vol. 2, Nonlinear Processes. Academic Press, New York 1967; 1971
3. Bellman, R.E.: Methods of Nonlinear Analysis, vol. 1, vol. 2, Academic Press, New York 1969, 1972
4. Beutelspacher, A., Petri, B.: Der Goldene Schnitt 2., überarbeitete und erweiterte Auflange. Elsevier GmbH, Spectrum Akademischer, Heidelberg 1996
5. Dunlap, R.A.: The Golden Ratio and Fibonacci Numbers (Original). World Scientific, Singapore 1977
6. Gelfand, I.M., Fomin, S.V.: Calculus of Variations. Prentice-Hall, Englewood Cliffs 1963
7. Iwamoto, S.: Theory of Dynamic Program: Japanese. Kyushu University Press, Fukuoka 1987

8. Iwamoto, S.: Cross dual on the Golden optimum solutions, Mathematical Economics, Kyoto University RIMS Koukyuroku, vol. 1443, pp 27–43, July 2005
9. Iwamoto, S.: The Golden optimum solution in quadratic programming. In: Takahashi, W., Tanaka, T. (eds) Proceedings of the International Conference on Nonlinear Analysis and Convex Analysis (Okinawa, 2005). Yokohama Publishers, Yokohama, pp. 199–205, 2007
10. Iwamoto, S.: The Golden trinity – optimility, inequality, identity. Mathematical Economics, Kyoto University RIMS Koukyuroku, vol. 1488, pp 1–14, May 2006
11. Iwamoto, S., Yasuda, M.: Dynamic programming creates the Golden Ratio, too. In: Proceedings of the 6th International Conference on Optimization: Techniques and Applications (ICOTA 2004), Ballarat, Australia, December 2004
12. Phelps, E.S.: The Golden rule of accumulation: A fable for growthmen. Am. Econ. Rev. **51**, 638–643 (1961)
13. Walser, H.: Der Goldene Schnitt. B.G. Teubner, Leibzig 1996

Adv. Math. Econ. 10, 91–100 (2007)

Advances in
MATHEMATICAL
ECONOMICS

A remark on law invariant convex risk measures

Shigeo Kusuoka*

Graduate School of Mathematical Sciences, The University of Tokyo, Komaba 3-8-1, Meguro-ku, Tokyo 153-8914, Japan
(e-mail: kusuoka@ms.u-tokyo.ac.jp)

Received: September 11, 2006
Revised: November 22, 2006

JEL classification: C65, G19

Mathematics Subject Classification (2000): 60B05

Abstract. The author gives a simple proof of the representation theorem for law invariant convex risk measures which was obtained by Kusuoka (Adv.Math.Econ. 3:83–95,2001), Frittelli and Rossaza Gianin (Adv.Math.Econ. 7:33–46,2005) and Jouini *et al.* (Adv.Math.Econ. 9:49–71,2006).

Key words: risk measure, law invariant.

1. Introduction

The idea of coherent risk measures has been introduced by Artzner *et al.*[1]. Then Föllmer and Scheid [3] extended this notion to convex risk measures. Let me introduce the definition of convex risk measures first.

Let $(\Omega, \mathcal{F}, P)$ be a probability space. We denote $L^\infty(\Omega, \mathcal{F}, P)$ by L^∞.

Definition 1. *We say that a map $\rho : L^\infty \to \mathbf{R}$ is a convex risk measure if the following are satisfied.*

1. $\rho(0) = 0$.
2. *For any $c \in \mathbf{R}$ and $X \in L^\infty$, we have*

$$\rho(X + c) = \rho(X) - c.$$

3. *If $X \geqq Y$, $X, Y \in L^\infty$, then $\rho(X) \leqq \rho(Y)$.*

* Partly supported by the 21st century COE program at Graduate School of Mathematical Sciences, The University of Tokyo.

4. *For any* $\lambda \in [0, 1]$, *and* $X, Y \in L^\infty$,

$$\rho(\lambda X + (1-\lambda)Y) \leqq \lambda\rho(X) + (1-\lambda)\rho(Y).$$

Remark 2. From the conditions (2) and (3) in Definition 1, we see that a convex risk measure $\rho : L^\infty \to \mathbf{R}$ satisfies

$$|\rho(X) - \rho(Y)| \leqq ||X - Y||_\infty, \quad X, Y \in L^\infty,$$

and so is continuous.

Also, we introduce the following notion.

Definition 3. *We say that a convex risk measure* $\rho : L^\infty \to \mathbf{R}$ *is law invariant, if* $\rho(X) = \rho(Y)$ *for any* $X, Y \in L^\infty$ *with the same probability laws.*

Let $\mathcal{D}$ be the set of probability distribution functions of bounded random variables, i.e., $\mathcal{D}$ is the set of non-decreasing right-continuous functions F on $\mathbf{R}$ such that there are $z_0, z_1 \in \mathbf{R}$ for which $F(z) = 0$, $z < z_0$, and $F(z) = 1$, $z \geqq z_1$. Let us define $Z : [0, 1) \times \mathcal{D} \to \mathbf{R}$ by

$$Z(x, F) = \inf\{z; F(z) > x\}, \quad x \in [0, 1), \quad F \in \mathcal{D}.$$

$Z(x, F)$ is a version of $F^{-1}(x)$. $Z(\cdot, F) : [0, 1) \to \mathbf{R}$ is a non-decreasing and right continuous function, and the probability distribution function of $Z(x, F)$ under the Lebesgue measure dx on $[0, 1)$ is F. We denote by F_X the probability distribution function of a random variable X.

For each $\alpha \in (0, 1]$, let $\rho_\alpha : L^\infty \to \mathbf{R}$ be given by

$$\rho_\alpha(X) = -\alpha^{-1} \int_0^\alpha Z(x, F_X)dx, \quad X \in L^\infty.$$

Also, we define $\rho_0 : L^\infty \to \mathbf{R}$ by

$$\rho_0(X) = -Z(0, F_X) = -\text{ess.}\inf X \quad X \in L^\infty.$$

Then it is easy to see that $\rho(X) : [0, 1] \to \mathbf{R}$ is a non-increasing continuous function for any $X \in L^\infty$.

Let $\mathcal{M}_{[0,1]}$ be the set of probability measures on $[0, 1]$.

Then combining the results by [5], Frittelli and Rossaza Gianin et al. [2] and Jouini et al. [4], we have the following.

Theorem 4. *Assume that* $(\Omega, \mathcal{F}, P)$ *is a standard atomless probability space. Let* $\rho : L^\infty \to \mathbf{R}$. *Then the following conditions are equivalent.*

1. *There is a subset $\mathcal{A}$ of the set $\mathcal{M}_{[0,1]} \times \mathbf{R}$ such that*

$$\sup\{b;\ (m,b) \in \mathcal{A}\} = 0$$

and

$$\rho(X) = \sup\left\{\int_{[0,1]} \rho_\alpha(X) m(d\alpha) + b;\ (m,b) \in \mathcal{A}\right\}, \quad X \in L^\infty.$$

2. *ρ is a law invariant convex risk measure.*

Our purpose of the present paper is to give a simple and direct proof for this theorem.

Remark 5. One can easily prove that

$$\rho_\alpha(X) = -\inf\left\{E[gX];\ g \in L^\infty,\ 0 \leqq g \leqq \frac{1}{\alpha},\ E[g] = 1\right\}, \quad X \in L^\infty,$$

for any $\alpha \in (0,1]$. Here we do not have to assume that $(\Omega, \mathcal{F}, P)$ is atomless. So we can easily check that ρ_α, $\alpha \in [0,1]$, are law invariant convex risk measures. Therefore, it is easy to prove that the condition (1) implies the condition (2) in Theorem 4.

2. Preparations

Let $N \geqq 2$. In this section, we consider a probability space $(\Omega_N, \mathcal{G}_N, P_N)$ such that $\Omega_N = \{1, \ldots, N\}$, $\mathcal{G}_N$ be the set of subsets of Ω_N, and $P_N(\{\omega\}) = \frac{1}{N}$, $\omega \in \Omega_N$.

Our aim in this section is to prove the following.

Theorem 6. *Let $\rho : L^\infty \to \mathbf{R}$. Then the following conditions are equivalent.*

1. *There is a subset $\mathcal{A}_0$ of the set $\mathcal{M}_{[0,1]} \times \mathbf{R}$ such that*

$$\sup\{b;\ (m,b) \in \mathcal{A}_0\} = 0$$

and

$$\rho(X) = \sup\left\{\int_{[0,1]} \rho_\alpha(X) m(d\alpha) + b,\ (m,b) \in \mathcal{A}_0\right\}, \quad X \in L^\infty.$$

2. *ρ is a law invariant convex risk measure.*

By Remark 4, it is sufficient to prove that the condition (2) implies the condition (1). So we prove the converse. Let ρ is a law invariant convex risk measure and let $\mathcal{C}$ be a subset of $L^\infty \times \mathbf{R}$ given by

$$\mathcal{C} = \left\{ (a,b) \in L^\infty \times \mathbf{R};\ \rho(X) \geqq -\sum_{i=1}^N a(i)X(i) + b, \quad \text{for all} \quad X \in L^\infty \right\}.$$

Since ρ is a convex function defined in L^∞ and L^∞ is finite dimensional, we see that

$$\rho(X) = \sup\left\{ -\sum_{i=1}^N a(i)X(i) + b;\ (a,b) \in \mathcal{C} \right\}, \qquad X \in L^\infty. \tag{1}$$

Moreover, we have the following.

Proposition 7. *For any* $(a,b) \in \mathcal{C}$, *we have the following.*

1. $a(i) \geqq 0, \quad i = 1, \dots, N.$
2. $\sum_{i=1}^N a(i) = 1.$
3. $\sup\{b;\ (a,b) \in \mathcal{C}\} = 0.$

Proof. Let $e_i \in L^\infty$, $i = 1, \dots, N$, such that $e_i(i) = 1$, and $e_i(j) = 0$, $j \neq i$. Then we have for any $c > 0$

$$0 \leqq -c^{-1}\rho(ce_i) \leqq a(i) - c^{-1}b.$$

Letting $c \to \infty$, we have the assertion (1).

Note that for any $c \in \mathbf{R}$, we have

$$0 = -\rho(c) - c \leqq c\left(\sum_{i=1}^N a(i) - 1\right) - b.$$

So we have for any $c > 0$

$$\left(\sum_{i=1}^N a(i) - 1\right) - c^{-1}b \leqq 0 \quad \text{and} \quad \left(\sum_{i=1}^N a(i) - 1\right) + c^{-1}b \geqq 0.$$

Letting $c \to \infty$, we have the assertion (2).

The assertion (3) is obvious, since $\rho(0) = 0$. □

Let $\mathcal{S}_N$ be the set of permutations on Ω_N. Then for any $a \in L^\infty$, there is a $\sigma_a \in \mathcal{S}_N$ such that

$$a(\sigma_a(N)) \leqq a(\sigma_a(N-1)) \leqq \cdots \leqq a(\sigma_a(1))$$

Then we have the following.

Proposition 8. 1. *For any* $(a, b) \in \mathcal{C}$, *and* $\sigma \in \mathcal{S}_N$, $(a \circ \sigma, b) \in \mathcal{C}$.
2. *Let* $(a, b) \in \mathcal{C}$ *and let* m_a *be a measure on* $[0, 1]$ *be given by*

$$\begin{aligned} m_a\left(\left\{\frac{j}{N}\right\}\right) &= (a(\sigma_a(j)) - a(\sigma_a(j+1)))j, \quad j = 1, \ldots, N-1, \\ m_a(\{1\}) &= a(\sigma_a(N))N \quad \textit{and} \quad m_a\left([0, 1] \setminus \left\{\frac{1}{N}, \frac{2}{N}, \ldots, 1\right\}\right) = 0. \end{aligned}$$

Then $m_a \in \mathcal{M}_{[0,1]}$ *and*

$$\max\left\{-\sum_{i=1}^{N}(a \circ \sigma)(i)X(i);\ \sigma \in \mathcal{S}_N\right\} = \int_{[0,1]} \rho_\alpha(X)m_a(d\alpha), \quad X \in L^\infty.$$

Proof. Let $X \in L^\infty$. Then it is obvious that random variables X and $X \circ \sigma^{-1}$ has the same probability law. Therefore, we have

$$\rho(X) = \rho(X \circ \sigma^{-1}) \geqq -\sum_{i=1}^{N} a(i)X(\sigma^{-1}(i)) + b = -\sum_{i=1}^{N} a(\sigma(i))X(i) + b.$$

This implies the assertion (1).

Now we will prove the assertion (2). Let $X \in L^\infty$. Then there is $\tau_X \in \mathcal{S}_N$ such that

$$X(\tau_X(1)) \leqq X(\tau_X(2)) \leqq \cdots \leqq X(\tau_X(N)).$$

It is easy to see that

$$X(\tau_X(k)) = N \int_{(k-1)/N}^{k/N} Z(x; F_X)dx, \quad k = 1, \ldots, N,$$

and so

$$\sum_{j=1}^{k} X(\tau_X(j)) = -k\rho_{k/N}(X), \quad k = 1, \ldots, N.$$

Then we have

$$\begin{aligned} \sum_{i=1}^{N} a(i)X(i) &= \sum_{i=1}^{N} a(\sigma_a(i))X(\sigma_a(i)) \\ &= \sum_{i=1}^{N} (a(\sigma_a(N)) + a(\sigma_a(i)) - a(\sigma_a(N)))X(\sigma_a(i)) \end{aligned}$$

$$
\begin{aligned}
&= a(\sigma_a(N))\left(\sum_{i=1}^{N} X(\sigma_a(i))\right) \\
&\quad + \sum_{i=1}^{N-1}\left(\sum_{j=i+1}^{N} (a(\sigma_a(j-1)) - a(\sigma_a(j)))X(\sigma_a(i))\right) \\
&= a(\sigma_a(N))\left(\sum_{i=1}^{N} X(\sigma_a(i))\right) \\
&\quad + \sum_{j=2}^{N}\left(\sum_{i=1}^{j-1} X(\sigma_a(i)))(a(\sigma_a(j-1)) - a(\sigma_a(j))\right) \\
&\geqq a(\sigma_a(N))\left(\sum_{i=1}^{N} X(\tau_X(i))\right) \\
&\quad + \sum_{j=2}^{N}\left(\sum_{i=1}^{j-1} X(\tau_X(i)))(a(\sigma_a(j-1)) - a(\sigma_a(j))\right) \\
&= -\int_{[0,1]} \rho_\alpha(X) m_a(d\alpha).
\end{aligned}
$$

Note that

$$
\sum_{i=1}^{N} a(\sigma_a \circ \tau_X^{-1}(i))X(i) = \sum_{i=1}^{N} a(i)X(\tau_X \circ \sigma_a^{-1}(i)) = -\int_{[0,1]} \rho_\alpha(X) m_a(d\alpha).
$$

So letting $X = 1$, we see that $m_a([0,1]) = 1$. These also imply the assertion (2). □

Now let

$$
\mathcal{A}_0 = \{(m_a, b) \in \mathcal{M}_{[0,1]} \times \mathbf{R};\ (a,b) \in \mathcal{C}\}
$$

Then we see from equation (1) and Proposition 7, that the condition (1) is satisfied for this $\mathcal{A}_0$. This completes the proof of Theorem 6.

3. Proof of Theorem 4

By Remark 5, it is sufficient to prove that the condition (2) implies the condition (1).

Let ρ be a law invariant convex risk measure, and let

$$\mathcal{A} = \Big\{(m,b) \in \mathcal{M}_{[0,1]} \times \mathbf{R}\,;\ \rho(X) \geqq \int_{[0,1]} \rho_\alpha(X) m(d\alpha) + b,\quad \text{for all } X \in L^\infty(\Omega)\Big\}.$$

Then it is sufficient to prove the following.

$$\rho(X) \leqq \sup\left\{\int_{[0,1]} \rho_\alpha(X) m(d\alpha) + b;\ (m,b) \in \mathcal{A}\right\}. \tag{2}$$

Since $(\Omega, \mathcal{F}, P)$ is atomless standard probability space, we may think that $\Omega = [0, 1)$, $\mathcal{F} = \mathcal{B}([0, 1))$, and P is a Lebesgue measure on $[0, 1)$. For any $n \geqq 1$, let

$$\mathcal{F}_n = \sigma\left\{1_{[(k-1)2^{-n}, k2^{-n})};\ k = 1, 2, \dots, 2^n\right\}.$$

Then we see that

$$\mathcal{F}_1 \subset \mathcal{F}_2 \subset \mathcal{F}_3 \subset \dots \quad \text{and} \quad \sigma\left(\bigcup_{n=1}^\infty \mathcal{F}_n\right) = \mathcal{F}.$$

Let

$$\mathcal{A}_n = \Big\{(m,b) \in \mathcal{M}_{[0,1]} \times \mathbf{R};\ \rho(X) \geqq \int_{[0,1]} \rho_\alpha(X) m(d\alpha) + b,\quad \text{for all } X \in L^\infty(\Omega, \mathcal{F}_n, P)\Big\}.$$

Then we have

$$\mathcal{A}_1 \supset \mathcal{A}_2 \supset \mathcal{A}_3 \supset \dots \supset \mathcal{A}.$$

Note that $\mathcal{M}_{[0,1]}$ is a compact subset of the dual space of $C([0, 1]; \mathbf{R})$ with weak * topology. Since $\rho(X) : [0, 1] \longrightarrow \mathbf{R}$ is continuous for all $X \in L^\infty$, $\mathcal{A}$, $\mathcal{A}_n$, $n = 1, 2, \dots,$ are closed in $\mathcal{M}_{[0,1]} \times \mathbf{R}$.

Proposition 9. *Let $\mathcal{A}_\infty = \bigcap_{n=1}^\infty \mathcal{A}_n$. Then $\mathcal{A}_\infty = \mathcal{A}$.*

Proof. Let $(m, b) \in \mathcal{A}_\infty$. Let $X \in L^\infty(\Omega, \mathcal{F}, P)$, and fix it. Let $Y \in L^\infty$ be given by $Y(\omega) = Z(\omega; F_X)$, $\omega \in \Omega = [0, 1)$. Since random variables X and Y have the same probability law, we see that $\rho(X) = \rho(Y)$. Let Y_n, $n = 1, 2, \dots,$ be random variables given by

$$Y_n(\omega) = Z\left(\frac{k}{2^n}-;\ F_X\right),\quad \frac{k-1}{2^n} \leqq \omega < \frac{k}{2^n},\quad k = 1, 2, \dots, 2^n.$$

Then we see that $Y_n(\omega) \downarrow Y(\omega)$, for any $\omega \in \Omega$. Since $(m, b) \in \mathcal{A}_n$, $n \geq 1$, we have

$$\rho(Y) \geqq \rho(Y_n) \geqq \int_{[0,1]} \rho_\alpha(Y_n)m(d\alpha) + b, \quad \alpha \in [0, 1].$$

It is easy to see that $\rho_\alpha(Y_n) \uparrow \rho_\alpha(Y)$, and so we have

$$\rho(X) = \rho(Y) \geqq \int_{[0,1]} \rho_\alpha(Y)m(d\alpha) + b = \int_{[0,1]} \rho_\alpha(X)m(d\alpha) + b.$$

This implies that $\mathcal{A}_\infty \subset \mathcal{A}$. It is obvious that $\mathcal{A}_\infty \supset \mathcal{A}$, and so we have the assertion. □

Now let us prove Theorem 4. For each $W \in L^\infty(\Omega_{2^n}, \mathcal{G}_{2^n}, P_{2^n})$, let $U_n(W) : \Omega \to \mathbf{R}$ be given by

$$U_n(W)(\omega) = \sum_{k=1}^{2^n} W(k)1_{[(k-1)2^{-n}, k2^{-n})}(\omega).$$

Then $U_n : L^\infty(\Omega_{2^n}, \mathcal{G}_{2^n}, P_{2^n}) \to L^\infty(\Omega, \mathcal{F}_n, P)$ is bijective. Let $\rho_n : L^\infty(\Omega_{2^n}, \mathcal{G}_{2^n}, P_{2^n}) \to \mathbf{R}$ be defined by $\rho_n(W) = \rho(U_n(W))$. Then it is easy to see that ρ_n is law invariant, convex risk measure and that

$$\rho_n(W) \geqq \int_{[0,1]} \rho_\alpha(W)m(d\alpha) + b, \quad W \in L^\infty(\Omega_{2^n}, \mathcal{G}_{2^n}, P_{2^n}),$$

if and only if

$$\rho(X) \geqq \int_{[0,1]} \rho_\alpha(X)m(d\alpha) + b, \quad X \in L^\infty(\Omega, \mathcal{F}_n, P),$$

for any $(m, b) \in \mathcal{M}_{[0,1]} \times \mathbf{R}$. This observation and Theorem 6 show that

$$\rho(X) = \sup\left\{\int_{[0,1]} \rho_\alpha(X)m(d\alpha) + b;\ (m, b) \in \mathcal{A}_n\right\}, \quad X \in L^\infty(\Omega, \mathcal{F}_n, P). \tag{3}$$

Let us take an arbitrary $X \in L^\infty(\Omega, \mathcal{F}, P)$ and fix it. Let Y and $\tilde{Y}_n$, $n = 1, 2, \ldots$, be random variables given by $Y(\omega) = Z(\omega, F_X)$, $\omega \in [0, 1)$, and

$$\tilde{Y}_n(\omega) = Z\left(\frac{k-2}{2^n} \vee 0;\ F_X\right), \quad \frac{k-1}{2^n} \leqq \omega < \frac{k}{2^n}, \ k = 1, 2, \ldots, 2^n.$$

Then we see that

$$Z(x;\ F_{\tilde{Y}_n}) = \tilde{Y}_n(x) \uparrow Y(x-), \quad x \in (0, 1),$$

and so we see that

$$\rho_\alpha(\tilde{Y}_n) \downarrow \rho_\alpha(Y) = \rho_\alpha(X), \quad n \to \infty, \ \alpha \in (0, 1].$$

Also, we see that

$$\rho_0(\tilde{Y}_n) = -\tilde{Y}_n(0) = -Y(0) = \rho_0(X)$$

So we see that $\rho_\alpha(\tilde{Y}_n)$ converges to $\rho_\alpha(X)$ uniformly in $\alpha \in [0, 1]$.

Since $\tilde{Y}_n \in L^\infty(\Omega, \mathcal{F}_n, P)$, we see from equation (3) that there exists $(m_n, b_n) \in \mathcal{A}_n$, for each $n \geq 1$, such that

$$\rho(\tilde{Y}_n) \leqq \int_{[0,1]} \rho_\alpha(\tilde{Y}_n) m_n(d\alpha) + b_n + \frac{1}{n}.$$

Note that

$$0 = \rho(0) \geqq \int_{[0,1]} \rho_\alpha(0) m_n(d\alpha) + b_n = b_n$$

and that

$$\begin{aligned} -\|\tilde{Y}_n\|_\infty &= \rho(\|\tilde{Y}_n\|_\infty) \leqq \rho(\tilde{Y}_n) \leqq \int_{[0,1]} \rho_\alpha(\tilde{Y}_n) m_n(d\alpha) \\ &\quad + b_n + \frac{1}{n} \leqq \|\tilde{Y}_n\|_\infty + b_n + 1. \end{aligned}$$

So we have

$$0 \geqq b_n \geqq -2\|\tilde{Y}_n\|_\infty - 1 \geqq -2\|X\|_\infty + 1.$$

Since $\mathcal{M}_{[0,1]}$ is compact, there are a subsequence $\{n_k; k = 1, 2, \ldots\}$ and $(m, b) \in \mathcal{M}_{[0,1]} \times \mathbf{R}$ such that

$$(m_{n_k}, b_{n_k}) \to (m, b), \quad n \to \infty, \quad \text{in } \mathcal{M}_{[0,1]} \times \mathbf{R}.$$

It is obvious that $(m, b) \in \mathcal{A}_{n_k}$, $k = 1, 2, \ldots$, and so we see that $(m, b) \in \mathcal{A}_\infty$. Also we have

$$\int_{[0,1]} \rho_\alpha(\tilde{Y}_{n_k}) m_{n_k}(d\alpha) \to \int_{[0,1]} \rho_\alpha(Y) m(d\alpha).$$

On the other hand, we see that

$$\rho(\tilde{Y}_n) \geqq \rho(Y) = \rho(X).$$

So we see that

$$\rho(X) \leqq \int_{[0,1]} \rho_\alpha(X) m(d\alpha) + b.$$

This proves that

$$\rho(X) \leqq \sup \left\{ \int_{[0,1]} \rho_\alpha(X) m(d\alpha) + b; \ (m, b) \in \mathcal{A} \right\}.$$

This completes the proof of Theorem 4.

References

1. Artzner, Ph., Delbaen, F., Eber, J.-M., Heath, D.: Coherent measures of risk. Math. Finance **9**, 203–228 (1999)
2. Frittelli, M., Rossaza Gianin, E.: Law invariant cobvex risk measures. Adv. Math. Econ. **7**, 33–46 (2005)
3. Föllmer, H., Scheid, A.: Convex measures of risk and trading constraints. Finance stochastics **6**, 429–447 (2002)
4. Jouini, E., Schachermayer, W., Touzi, N.: Law invariant risk measures have the Fatou property. Adv. Math. Econ. **9**, 49–71 (2006)
5. Kusuoka, S.: On law invariant coherent risk measures. Adv. Math. Econ. **3**, 83–95 (2001)

Adv. Math. Econ. 10, 101–119 (2007)

Advances in
MATHEMATICAL
ECONOMICS

Existence and uniqueness of an equilibrium in a model of spatial electoral competition with entry*

Anna Rubinchik[1] **and Shlomo Weber**[2,3,4]

[1] Department of Economics, University of Colorado at Boulder, UCB 256, Boulder, CO 80309, USA
(e-mail: Anna.Rubinchik@colorado.edu)
[2] SMU, Department of Economics, Dallas, TX, USA
(e-mail: sweber@mail.smu.edu)
[3] CORE, Catholic University of Louvain, Louvain, Belgium
[4] CEPR, London, UK

Received: October 3, 2006
Revised: October 3, 2006

JEL classification: C62, C72, D72

Mathematics Subject Classification (2000): 91A10, 91B12

Abstract. Two incumbent parties choose their platforms in a unidimensional policy space while facing a credible threat of an entry by the third party. Relative electoral support is the predominant objective of each party, and the third party enters only if it can displace one of the incumbents by receiving at least the second highest support. We prove the existence and uniqueness of an equilibrium for a wide class of distributions of voters' ideal points, including, in particular, log-concave distribution functions. Moreover, in an equilibrium the incumbents prevent the entry and achieve the "balance of power" by choosing distinct positions in the policy space and equally splitting the electorate.

Key words: incumbent parties, threat of entry, entry-deterrence, rank concerns, balance of power

* We would like to thank Michel Le Breton and Jean-François Mertens for their useful comments. A part of the paper was written when the first author was visiting CORE, the hospitality of which is greatly appreciated.

1. Introduction

We consider a model of spatial competition with two incumbents and a potential entrant, where parties are pre-occupied by their rank (or relative performance).[1] Competing parties choose their positions in the unidimensional policy space, and each voter supports the party that proposes a platform closest to her ideal point.[2] Similar to Palfrey [8], the incumbents in this model behave as Nash players with respect to one another, but as Stackelberg leaders with respect to the entrant. In other words, the incumbents choose their platforms simultaneously, but in full anticipation of the third party's location along the spectrum of political issues. For entry to be deemed successful, the third party must garner more votes than at least one of the established parties; otherwise, the potential entrant stays out of the race altogether. Thus, becoming one of the top two is crucial for a party.

The focus of the paper is to show existence and to characterize an equilibrium of this game, referred to as $\mathcal{R}$-equilibrium.

We find that under rather mild restrictions on the distribution of voters' ideal points, a third party entry is not sustainable in an $\mathcal{R}$-equilibrium, i.e., the potential entrant can never become one of the top two. This immediately implies that the set of $\mathcal{R}$-equilibria is a subset of incumbent strategies which prevent entry by a third party, a notion introduced by Greenberg and Shepsle [7]. This is true even though in our setting the incumbents are forward looking; in particular, each of them might want to trigger an entry to displace the other incumbent from being among the top two, thereby improving her own rank or increasing the share of own supporters, while preserving its relative standing.

Greenberg and Shepsle [7] pointed out that a profile of entry-preventing incumbents' strategies, which we refer to as $\mathcal{D}$-strategies, may fail to exist. We derive quite general sufficient conditions for existence and uniqueness of $\mathcal{D}$-strategies. We go on to demonstrate that $\mathcal{R}$-equilibrium exists under these conditions if, and only if incumbents achieve a balance of power whereby the electorate is shared equally among them.

The paper is organized as follows. The following section describes the spatial competition game. Section 3 demonstrates that an equilibrium of the game has to be entry-deterrent. Section 4 contains concluding remarks.

---

[1] Although in what follows we refer to the competitors as political parties, the model is, clearly, applicable to other environments, e.g., an oligopolistic competition with differentiated products.

[2] One could view this voting behavior as a desire to associate oneself with a certain platform, defining one's identity. This motif might prevail in large elections, in which pivotalness of an individual voter is minuscule. See [6] for a related overview. Moreover, there are reasons to believe that sincere voting is a good "positive" assumption in the view of the recent empirical findings, see, for example, [1].

2. The model

Consider two incumbent parties, 1 and 2, and one potential entrant, e. The competing parties choose their positions in the issue space $I = [0, 1]$. Having observed their choices, the entrant can either stay out (the option denoted by N) or enter the race and choose a platform in I.

Payoffs to the parties are based on voters' support. Each voter has symmetric single-peaked preferences with the most preferred alternative, or *ideal point*, in I. Let F be the cumulative distribution of the voters' ideal points in the issue space I, with $F(0) = 0$ and $F(1) = 1$.

Given positions of the competing parties, each voter supports the party whose position is closest to her ideal point, and she randomly picks one of the closest by, in case there are many. No abstention is allowed — each voter identifies herself with a party. Let $x = (x_1, x_2)$ be a pair of positions chosen by the established parties 1 and 2 and assume that $x_1 \leq x_2$, let $X = (x, x_e)$ be the choices made by all three parties, where the entrant can either choose a platform or decide not to enter: $x_e \in I \cup \{N\}$.

All three parties have lexicographic preferences. Each of them first considers the rank and then the fraction of votes they garner. Given choice X of the parties, we denote by $r_1(X)$, $r_2(X)$, $r_e(X)$ their corresponding ranks. Rank $r_i(X)$ can obtain one of the six values:

- $A1$ the sole possession of the first place,
- $A2$ sharing the first place with one of the other parties,
- $B1$ the sole possession of the second place (with only one party being ahead),
- $A3$ sharing the first place with two other parties,
- $B2$ sharing the second place with one of the other parties,
- $C1$ the sole possession of the third place.

We assume the following natural preferences for all parties:

$$A1 \succ A2 \succ B1 \succ A3 \succ B2 \succ C1.$$

If the entrant decides to stay out, we say that $r_e(X) = N$, and her ranking reflects the desire to enter only if she can become at least the second (attaining rank $B1$) in the electoral competition.[3] That is, the preferences of the entrant are given by

$$A1 \succ A2 \succ B1 \succ N \succ A3 \succ B2 \succ C1.$$

[3] As mentioned in the introduction, in many situations "winning" is associated with being one of the top two and if entry is sufficiently costly, it will be prevented in case the entrant is not expecting to "win". An alternative specification in which the entrant still enters if she can assure sharing the first place with the incumbents will lead to the same results, as the outcome $A3$ can not be supported in $\mathcal{R}$-equilibrium.

If two different outcomes generate the same rank, a party prefers the one that yields a higher vote share. This share is also determined by the positions of the three parties.

Suppose first that party e does not enter. If the incumbents choose the same position in the issue space, each party is supported by one half of the electorate. If their positions are different, $x_1 < x_2$, then party 1 is supported by all those voters whose ideal points are to the left of the middle point $\frac{x_1+x_2}{2}$, whereas party 2 is supported by voters whose ideal points are to the right of $\frac{x_1+x_2}{2}$. That is, the support of party $i = 1, 2$ denoted by $s_i(X)$ is determined by

$$s_1(X) = F\left(\frac{x_1 + x_2}{2}\right),$$

$$s_2(X) = 1 - F\left(\frac{x_1 + x_2}{2}\right).$$

Suppose now that party e enters and chooses position x_e in the issue space. If all three parties choose the same position, then each is supported by one-third of the electorate, i.e., $s_1(X) = s_2(X) = s_e(X) = \frac{1}{3}$. If the incumbents choose the same position but the entrant locates herself at a different point, say at $x_1 = x_2 < x_e$, then the support of each incumbent is given by $F(\frac{x_1+x_e}{2})/2$, whereas party e is supported by $1 - F(\frac{x_1+x_e}{2})$ voters. If all parties choose different positions, say $x_1 < x_2 < x_e$, then the support of parties 1, 2 and e is given by $F(\frac{x_1+x_2}{2})$, $F(\frac{x_2+x_e}{2}) - F(\frac{x_1+x_2}{2})$ and $1 - F(\frac{x_2+x_e}{2})$, respectively. That is, if party i has rivals both to the left and to the right,[4] then its support covers the interval between two points: one equidistant from x_i and its opponent from the left side, and the other equidistant from x_i and its opponent from the right. If party has no rivals to its left, then its support covers the interval between 0 and the point which is equidistant from x_i and its closest opponent from the right side. Similarly, if party i has no rivals to its right,[5] then its support covers the interval between the point which is equidistant from x_i and its closest opponent from the left side and the right endpoint of the issue space.

Unfortunately, subgame perfect equilibria of the game described above do not always exist. Let us first examine the reasons for the nonexistence and then modify the game to avoid the problem.

Recall party e enters only when it can win higher support than at least one of the incumbents. To describe these cases, fix the positions of the incumbent parties, $x = (x_1, x_2)$, and consider the following sets:

$$D_1(x) = \{x_e \in I | s_e(x, x_e) > \max[s_1(x, x_e), s_2(x, x_e)]\}$$

[4] Eaton and Lipsey [4] call this location *interior*.

[5] A position which has no rivals either to the left or to the right is called *peripheral* in [4].

or the set of positions for the entrant that guarantee the entrant the sole possession of the first place;

$$D_{12}(x) = \{x_e \in I | s_e(x, x_e) = \max[s_1(x, x_e), s_2(x, x_e)] \\ > \min[s_1(x, x_e), s_2(x, x_e)]\}$$

or the set of positions that yield the entrant the share of the first place with one of the incumbents; and

$$D_2(x) = \{x_e \in I | \max[s_1(x, x_e), s_2(x, x_e)] \\ > s_e(x, x_e) > \min[s_1(x, x_e), s_2(x, x_e)]\}$$

or the set of positions where the entrant holds the second place. Also, let

$$D(x) = D_1(x) \cup D_{12}(x) \cup D_2(x) \\ = \{x_e \in I | s_e(x, x_e) > \min[s_1(x, x_e), s_2(x, x_e)]\}.$$

Given the preferences of party e, it will not enter, if the set $D(x)$ is empty. Otherwise, it considers the sets $D_1(x)$, $D_{12}(x)$, $D_2(x)$ (in this order) and makes its vote-maximizing choice over the first nonempty set in this sequence.

By using the arguments of Palfrey [8], it is easy to see that the best response of the entrant over the sets $D_1(x)$ or $D_2(x)$ may fail to exist. (We shall show that the set $D_{12}(x)$ is either empty or consists of a unique element.) In this case, we adopt the procedure offered by Palfrey [8] and extended by Weber [10]. Namely, we consider the average of the incumbents' payoffs over the set of "ε-best" responses of the entrant. Specifically, for each incumbent i and for each positive ε we determine the average of player i's payoffs over the set of "ε-best" responses of the entrant over the set $D_1(x)$ if nonempty, and over D_2, otherwise, and consider its limit when ε approaches zero.

If the entrant chooses to enter, so that for a given pair of incumbents' positions, x, the set $D(x)$ is nonempty, define by $E(x)$ the subset of $D(x)$ that the entrant considers, i.e., let

$$E(x) = \begin{cases} D_1(x) & \text{if} \quad D_1(x) \neq \varnothing \\ D_{12}(x) & \text{if} \quad D_1(x) = \varnothing, \; D_{12}(x) \neq \varnothing \\ D_2(x) & \text{if} \quad D_1(x) \cup D_{12}(x) = \varnothing, \; D_2(x) \neq \varnothing. \end{cases}$$

For each pair of incumbent choices $x \in I^2$ and for $\varepsilon \geq 0$, denote by $B_\varepsilon^E(x)$ the set of ε-best responses over $E(x)$. That is,

$$B_\varepsilon^E(x) = \{x_e \in E(x) | s_e(x, x_e) \geq s_e(x, y) - \varepsilon \; for \; all \; y \in E(x)\}.$$

Let a pair of incumbents' strategies $x = (x_1, x_2)$ be such that the set $D(x)$ is nonempty, and moreover, $E(x) \neq D_{12}(x)$. While the set of best responses

$B_0^E(x)$ might be empty, the set $B_\varepsilon^E(x)$ is nonempty for every strictly positive ε.[6] Let

$$\mu_\varepsilon(x) \equiv \int_{B_\varepsilon^E(x)} dz \quad \text{and} \quad u_i(x) \equiv \lim_{\varepsilon \to 0} \int_{B_\varepsilon^E(x)} \frac{1}{\mu_\varepsilon(x)} s_i(x, z) dz$$

for each party i, $i = 1, 2$. Palfrey [8] has shown that the functions u_i are well defined. Roughly speaking, u_i is a limit of incumbent i's electoral support, provided that the potential entrant is mixing across his ε-best responses with equal probability.

Now we can define the second component of the preferences for the incumbent parties. If the best response of the entrant exists – either the entrant does not enter, so that $x_e = N$, or her vote-maximizing position given is well defined[7] – this component is s_i, as defined before, and uniqueness of the best reply of the entrant in this case implies x_e is a function of the positions of the incumbents $x = (x_1, x_2)$. If the set $B_0^E(x)$ is empty, then her payoff is set to be equal to the limit of the average support, u_i, which is also fully determined by the positions of the incumbents. To sum up, an incumbent i, $i = 1, 2$, derives payoff $\pi_i(x)$ from her electoral support:

$$\pi_i(x) \equiv \begin{cases} u_i(x), & \text{if } D(x) \neq \varnothing, \ \ B_0^E(x) = \varnothing \\ s_i(x, x_e(x)), & \text{otherwise.} \end{cases} \tag{1}$$

We can now formally define game Γ between the incumbents, who foresee that the third party, e, enters only if she can displace one of the incumbents and to guarantee at least the sole position of the second place. Formally,

Definition 2.1. *In the two-person game Γ the incumbents have strategy set I. Players' preferences are lexicographic in (1) rank, r_i and (2) payoff, π_i. A pure strategy equilibrium of the game Γ is called an $\mathcal{R}$-equilibrium.*

It is important to distinguish our equilibrium notion from that introduced by Greenberg Shepsle [7]. They refer to each pair of incumbent positions that prevent entry by a third party as 2-equilibrium. We simply call these strategies *entry-deterrent*:

Definition 2.2. *A pair of positions of established parties $x = (x_1, x_2)$ is called entry-deterrent ($\mathcal{D}$-*strategies*), if the set $D(x)$ is empty.*

Note that an entry-deterrent pair of incumbent strategies is not necessarily consistent with $\mathcal{R}$-equilibrium. Indeed, the latter requires the incumbents'

---

6 It can be shown, using the argument in [11] that the set $B_\varepsilon^E(x)$ is the union of a finite set of intervals and that if the set of best responses $B_0^E(x)$ is nonempty, it consists of a unique element (see claim 3.1 in [11]).

7 As is, for example, in case $E(x) = D_{12}(x)$.

positions to be immune to unilateral deviations by the incumbents, while correctly anticipating the response of a potential third party. Thus, we allow an incumbent to induce the entry, if it is in her interest. The next section offers a characterization of $\mathcal{R}$-equilibria. The key result is that in $\mathcal{R}$-equilibria neither of the incumbents will want to induce the entry.

3. Balance of power in $\mathcal{R}$-equilibrium

In this section, we derive conditions for the existence and uniqueness of both $\mathcal{R}$-equilibrium and $\mathcal{D}$-strategies, and study the relationship between the two. Throughout the remainder of the paper we consider only pairs of strategies (x_1, x_2) where party 1 is located to the left of party 2, i.e., $x_1 \leq x_2$, so that the uniqueness of an equilibrium will be stated in terms of equilibrium configurations up to a permutation of incumbents' strategies. The proofs of all results in this section are relegated to the Appendix.

We shall now introduce two assumptions, (A.1) and (A.2), which are maintained for the rest of the paper. The first is quite standard and requires the distribution of voters' ideal points to be unimodal and the density function f to be continuous:

Assumption (A.1). $f(\cdot)$ is continuous and strictly positive on $[0, 1]$. Moreover, there exists $\hat{x} \in I$, such that $f(\cdot)$ is strictly increasing on the interval $[0, \hat{x}]$ and strictly decreasing on the interval $[\hat{x}, 1]$.

The second assumption assures that the ideal points of the voters are not too concentrated at any given interval. Following Haimanko, Le Breton and Weber (2005) we will refer to this assumption as *gradually escalating median (GEM)*.[8] Let $l : [0, 1] \rightarrow [0, 1]$ be the median of $[0, t]$ and $r : [0, 1] \rightarrow [0, 1]$ be the median of $[t, 1]$ under F. Given the first assumption both functions are continuously differentiable.

Assumption (A.2). $l'(t) < 1, r'(t) < 1$.

Assumptions (A.1) and (A.2) will allow us to compare $\mathcal{D}$-strategies and strategies of the incumbents under $\mathcal{R}$-equilibria. The definition of $\mathcal{D}$-strategies rules out a move by the entrant, while $\mathcal{R}$-equilibrium requires an incumbent's position to be immune against a unilateral deviation of another incumbent that can "invite" an entry by the third party.

[8] This assumption is weaker than log-concavity, which is rather mild on its own, see [2] for the discussion and connection to other properties, including monotone hazard ratio. More precisely, a stronger assumption than (A.2) would require function $F(t)$ to be log-concave on the interval $[0, \hat{x}]$ and the function $1 - F(1 - t)$ to be log-concave on the interval $[\hat{x}, 1]$.

As defined, $\mathcal{R}$-equilibrium does not preclude an entrance by the third party. It is important to establish, therefore, whether there exists $\mathcal{R}$-equilibrium in which the established parties allow for the entry of party e. Proposition 3.1 demonstrates that the answer is negative, implying that the notion of $\mathcal{R}$-equilibrium is no less restrictive than the notion of entry-deterrent strategies.

Proposition 3.1. *Assume that (A.1) and (A.2) hold. Then in any $\mathcal{R}$-equilibrium party e does not enter. That is, every pair of incumbents' $\mathcal{R}$-equilibrium strategies is also a pair of $\mathcal{D}$-strategies.*

Greenberg and Shepsle [7] concluded that, in general, the set of $\mathcal{D}$-strategies might be empty. Providing sufficient conditions for existence of these strategies remained open. Cohen [3] has shown that $\mathcal{D}$-strategies exists for the special case where the distribution of voters' ideal points is given by a normal density function. The following proposition demonstrates that the condition of normality, and even symmetry, of the distribution can be dropped. Unimodality and GEM yield existence and uniqueness of $\mathcal{D}$-strategies.

Proposition 3.2. *Under (A.1) and (A.2), there is a unique pair of $\mathcal{D}$-strategies $x^d = (x_1^d, x_2^d)$. Moreover, x^d satisfies*[9]

$$F(x_1^d) = \frac{1}{2} F\left(\frac{x_1^d + x_2^d}{2}\right), \tag{2}$$

$$1 - F(x_2^d) = \frac{1}{2}\left(1 - F\left(\frac{x_1^d + x_2^d}{2}\right)\right), \tag{3}$$

$$\frac{1}{3} < F\left(\frac{x_1^d + x_2^d}{2}\right) < \frac{2}{3}. \tag{4}$$

Our next proposition derives necessary and sufficient conditions for existence of $\mathcal{R}$-equilibrium. Note that by Proposition 3.1, the set of $\mathcal{R}$-equilibria is a subset of the set of $\mathcal{D}$-strategies. We show that, in general, the converse is not true. Since, by Proposition 3.2, for a given distribution of ideal points a pair of $\mathcal{D}$-strategies is unique, it follows that the set of $\mathcal{R}$-equilibria might be empty. That is, even though [under (A.1) and (A.2)], there is always a unique entry-deterring pair strategies for established parties, one of the incumbents could be better-off by deviating from it, thus *allowing for entry* of party e. Given that we impose an additional requirement of Nash behavior on incumbents, it is not surprising to find out that $\mathcal{R}$-equilibrium may fail to exist in circumstances which guarantee existence of $\mathcal{D}$-strategies. Our result shows that $\mathcal{R}$-equilibrium exists only in the case where the established parties, while locating themselves

[9] These equations were formulated in Greenberg and Shepsle [7] as necessary conditions for the existence of a $\mathcal{GS}$-equilibrium.

at quartiles of the distribution, achieve a balance of power by equally splitting the total electoral vote.

Proposition 3.3. *Assume that (A.1) and (A.2) hold and let the pair* $x^d = (x_1^d, x_2^d)$ *be* $\mathcal{D}$*-strategies. Then a pair of incumbents' strategies* x^d *is an* $\mathcal{R}$*-equilibrium if and only if*

$$F\left(\frac{x_1^d + x_2^d}{2}\right) = \frac{1}{2}. \tag{5}$$

That is, if

$$F^{-1}\left(\frac{1}{4}\right) + F^{-1}\left(\frac{3}{4}\right) = 2F^{-1}\left(\frac{1}{2}\right), \tag{6}$$

then the set of $\mathcal{R}$*-equilibria consists of the unique element,* $(F^{-1}(\frac{1}{4}), F^{-1}(\frac{3}{4}))$.

Otherwise, the set of $\mathcal{R}$*-equilibria is empty.*

The intuition behind Proposition 3.3 is quite simple. When power is balanced between incumbents, in the sense that each garners 50% of the vote, neither party can improve its standing by altering its position. Take, for example, the candidate located on the left. Moving further to the left will reduce her support relative to the other incumbent as well as make it possible for the entrant to locate "very close" on her right and displace her by garnering slightly more than 25% of the vote. On the other hand, should the left-most incumbent attempt to increase her support by moving closer to the current rival, the entrant will displace her by locating "very close" on her left and garnering slightly more than 25% of the vote. By the same argument, the right-most candidate cannot improve her standing when the incumbents choose platforms in such a manner that the electorate is divided equally between the established parties.

Balance of power is essential for $\mathcal{R}$-equilibrium to exist. If power is not shared equally between the two incumbents, the second place incumbent can improve its standing by moving slightly closer to the incumbent who ranks first. In doing so, the entrant can now garner more votes than the top-ranked incumbent by entering "very close" on her outside. The incumbent formerly in second place will now win the election, the entrant will get the second place, and the incumbent formerly in first place will now be ranked third as it is "squeezed" between its old rival and the new third party.

Implicitly, Proposition 3.3 characterizes societies (described by distributions of voters' ideal points) for which $\mathcal{R}$-equilibrium exists. In particular, any symmetric density function satisfies condition (6). Clearly, symmetry is not necessary for that condition to hold.

4. Conclusions

The model offers an analysis of electoral competition in the presence of rank concerns. If the distribution of voters' ideal points F is single-peaked and satisfies the gradually-escalating-median property, incumbents' strategies in an $\mathcal{R}$-equilibrium are always entry-deterrent and in this case there exists a unique $\mathcal{R}$-equilibrium if and only if the distribution F satisfies

$$F^{-1}\left(\frac{1}{4}\right)+F^{-1}\left(\frac{3}{4}\right)=2F^{-1}\left(\frac{1}{2}\right).$$

Appendix

Let a pair of incumbents' strategies $x=(x_1,x_2)$ be given. Let $\alpha\equiv\frac{x_1+x_2}{2}$, $\mathbf{I}_1\equiv F(x_1)$, $\mathbf{I}_2\equiv F(\alpha)-F(x_1)$, $\mathbf{I}_3\equiv F(x_2)-F(\alpha)$, $\mathbf{I}_4\equiv 1-F(x_2)$, $G(x)\equiv 1-F(x)$.

Assume, without loss of generality, that

$$\mathbf{I}_1\geq\mathbf{I}_4. \tag{7}$$

Lemma 5.1. *If $\mathbf{I}_1\geq\mathbf{I}_2$ then the support of an entrant choosing a policy $z\in(x_1,x_2)$ will not exceed that of the first party and if $I_3\leq I_4$ the support of an entrant choosing a policy $z\in(x_1,x_2)$ will not exceed that of the second party. If both inequalities $\mathbf{I}_1\geq\mathbf{I}_2$ and $\mathbf{I}_3\leq\mathbf{I}_4$ hold, then no entry will occur between x_1 and x_2, i.e., the set*

$$D^m(x)\equiv\{z\in D(x)|x_1<z<x_2\}$$

is empty.

Proof. Consider the entry of party e between x_1 and x_2, say, at z. Let $\mathbf{I}_1\geq\mathbf{I}_2$ hold. Then the support of the entrant is equal to $F(z+r_2)-F(z-r_1)$ where $r_2(z)\equiv(x_2-z)/2$, $r_1(z)\equiv(z-x_1)/2$. This implies $r_1'(z)=-r_2'(z)=\frac{1}{2}$ and then by assumption $(A.2)$,

$$\begin{aligned}l(z+r_2)-l(\alpha)&<z+r_2-\alpha\\&=z-r_1-x_1,\end{aligned}$$

thus

$$l(z+r_2)<z-r_1,$$

as $l(\alpha)\leq x_1$. Then $F(z+r_2)<2F(z-r_1)$, so that the first party has higher support than the entrant. Similarly, if $\mathbf{I}_3\leq\mathbf{I}_4$, $G(1-z+r_1)<2G(1-z-r_2)$

and the second party has a higher support than the entrant. If both inequalities $\mathbf{I}_1 \geq \mathbf{I}_2$ and $\mathbf{I}_3 \leq \mathbf{I}_4$ hold,

$$F(z + r_2) - F(z - r_1) \leq \min[F(z - r_1), 1 - F(z + r_2)].$$

Hence, the support of the entrant does not exceed that of any of the two established parties. □

In order to prove Proposition 3.2 we shall use the following lemma:

Lemma 5.2. *For each* $y, 0 < y < 1$, *define the values of* $a(y)$ *and* $b(y)$ *by*

$$2F(a(y)) = F(y) \quad \text{and} \quad 2(1 - F(b(y))) = 1 - F(y). \tag{8}$$

Then there is a unique y^d, *satisfying*

$$a(y^d) + b(y^d) = 2y^d. \tag{9}$$

Moreover, the value of y^d *is such that* $\frac{1}{3} < F(y^d) < \frac{2}{3}$.

Proof. It is easy to verify that the functions $a(\cdot)$ and $b(\cdot)$ are well defined, increasing and differentiable on the interval [0, 1], and so is $h(y) \equiv 2y - a(y) - b(y)$. Since $a(0) = 0, b(0) > 0, a(1) < 1, b(1) = 1$, it follows that $h(0) = -b(0) < 0$ and $h(1) = 1 - a(1) > 0$. Hence, there exists a value y that solves (9).

For uniqueness rewrite the necessary conditions as a system of two equations,

$$\begin{cases} 2F(a) - F\left(\frac{a+b}{2}\right) = 0 \\ 2F(b) - F\left(\frac{a+b}{2}\right) = 1 \end{cases}$$

for $(a, b) \in [0, 1]^2$. Its Jacobian is

$$\begin{bmatrix} 2f(a) - \frac{1}{2}f\left(\frac{a+b}{2}\right) & -\frac{1}{2}f\left(\frac{a+b}{2}\right) \\ -\frac{1}{2}f\left(\frac{a+b}{2}\right) & 2f(b) - \frac{1}{2}f\left(\frac{a+b}{2}\right) \end{bmatrix},$$

the principal minors of which are positive. Indeed, by (A.2),

$$l'(x) = \frac{1}{2}\frac{f(x)}{f(l(x))} < 1 \tag{10}$$

$$r'(x) = \frac{1}{2}\frac{f(x)}{f(r(x))} < 1.$$

This leads to the following conditions:

$$2f(a) - \frac{1}{2}f\left(\frac{a+b}{2}\right) > \frac{1}{2}f\left(\frac{a+b}{2}\right) > 0 \qquad (11)$$
$$2f(b) - \frac{1}{2}f\left(\frac{a+b}{2}\right) > \frac{1}{2}f\left(\frac{a+b}{2}\right) > 0$$

and, therefore,

$$\left[2f(a) - \frac{1}{2}f\left(\frac{a+b}{2}\right)\right]\left[2f(b) - \frac{1}{2}f\left(\frac{a+b}{2}\right)\right] - \left[\frac{1}{2}f\left(\frac{a+b}{2}\right)\right]^2 > 0. \qquad (12)$$

Conditions (11) and (12) assure the positiveness of the principal minors of the Jacobian and, thus, uniqueness by the Fundamental Global Univalence Theorem ([9], p. 20).[10]

Finally, we need to verify the inequalities

$$\frac{1}{3} < F(y^d) < \frac{2}{3}.$$

By the necessary conditions (8), we know $l\left(y^d\right) = a$. By assumption (A.2) for any $x \in \left[y^d, b\right]$, $l'(x) < 1$, or for $r \in \left[0, b - y^d\right]$

$$l\left(y^d + r\right) - r < l\left(y^d\right),$$

which implies

$$l(b) < l\left(y^d\right) + b - y^d = y^d$$

thus

$$\frac{F(b(y^d))}{F(y^d)} < \frac{F(y^d)}{F(a(y^d))} = 2$$

and, moreover,

$$F(y^d) > F(b(y^d)) - F(y^d) = \frac{1 - F(y^d)}{2},$$

yielding $F(y^d) > \frac{1}{3}$. Similarly,

$$\frac{1 - F(a(y^d))}{1 - F(y^d)} < \frac{1 - F(y^d)}{1 - F(b(y^d))} = 2,$$

[10] The initial formulation is due to [5].

which implies

$$1 - F(y^d) > F(y^d) - F(a(y^d)) = \frac{F(y^d)}{2},$$

or $F(y^d) < \frac{2}{3}$. □

Before proceeding to Proposition 3.1, we provide a proof of Proposition 3.2.

Proof of Proposition 3.2. Greenberg and Shepsle [7] have shown that conditions (2), (3), and (4) are necessary for a pair of strategies to be $\mathcal{D}$-strategies. By Lemma 5.2, there is a unique pair of strategies which satisfies those conditions. Thus, it remains to show that this pair of strategies is indeed a pair of $\mathcal{D}$-strategies under (A.1) and (A.2).

Lemma 5.2 yields the existence of a real number $y \in (0, 1)$ such that the pair $w = (a(y), b(y))$ is the unique pair of strategies which satisfies (2), (3) and (4). To demonstrate that set $D(w)$ is empty, let us consider options available for party e, given the incumbents' locations at points $a(y)$ and $b(y)$, respectively. If party e enters to the left of $a(y)$, it would not displace party 1. Moreover, since $F(a(y)) \leq \frac{1}{3} \leq F(y)$, party 2 would not be displaced either. By using similar arguments one can show that neither of the established parties would be displaced if party e enters to the right of $b(y)$. In the view of conditions (2), (3) lemma 5.1 guarantees that no entrant will choose a position in between $a\,(y)$ and $b\,(y)$. □

We shall turn now to the proof of Proposition 3.1.

For each pair of positions of the established parties $x = (x_1, x_2)$ denote by $y^1 = y^1(\frac{x_2-x_1}{2})$ and $y^2 = y^2(\frac{x_2-x_1}{2})$ two locations in the issue space which satisfy

$$F(y^1 + \frac{x_2 - x_1}{2}) = 2F(y^1); \tag{13}$$

$$G(y^2 - \frac{x_2 - x_1}{2}) = 2G(y^2). \tag{14}$$

Let

$$z^1(x) \equiv 2y^1 - x_1; \quad z^2(x) \equiv 2y^2 - x_2.$$

In the case of $z^1(x) > x_1$, if party e enters at $z^1(x)$ it would generate the support equal to that of party 1. Assumption (A.2) can be used to show that if party e enters between x_1 and $z^1(x)$ it would displace party 1, and if it enters to the right of $z^1(x)$ it would not displace party 1. Similarly, in case $z^2(x) < x_2$, if party e enters to the left of $z^2(x)$ it would not displace party 2, if it enters at $z^2(x)$ it would generate the same number of votes as party 2, and if it enters between $z^2(x)$ and x_2 it would displace party 2. In addition, for each pair of positions of the established parties $x = (x_1, x_2)$, consider the function $s^e(x, t)$ which determines the support of party e generated by entry at $t \in I$. It has been shown in [10] that assumptions (A.1) guarantee that $s^e(x, \cdot)$ is continuous and strictly quasi-concave on the interval (x_1, x_2). It allows us to continuously extend this function to the closed interval $[x_1, x_2]$. Thus, there exists a unique value $z^\star(x)$, $x_1 \le z^\star(x) \le x_2$ such that

$$z^\star(x) = \arg \max_{x_1 \le t \le x_2} s^e(x, t).$$

$z^\star(x)$ determines the vote-maximizing location[11] of party e between the positions of the two established parties.

We will show now that if one of the sets $D_1(x)$, $D_{12}(x)$ or $D_2(x)$ is nonempty then x is not an $\mathcal{R}$-equilibrium.

Lemma 5.3. *Let a pair of strategies $x = (x_1, x_2)$ be an $\mathcal{R}$-equilibrium. Then set $D(x) \backslash D_2(x) = D_1(x) \cup D_{12}(x)$ is empty.*

Proof. Let $x = (x_1, x_2)$ be an $\mathcal{R}$-equilibrium. Suppose first that $D_1(x)$ is nonempty. Since by assumption (7), $\mathbf{I}_1 \ge \mathbf{I}_4$, party e may enter either to the left of x_1 or between x_1 and x_2.

(i). Suppose that party e enters to the left of x_1, which will happen, only if $\mathbf{I}_1 > \mathbf{I}_4$. The entrant gains the support of almost $\mathbf{I}_1$ and wins the race, if $\mathbf{I}_1 > \mathbf{I}_4 + \mathbf{I}_3$. But then a move of player 2 to the left of x_2 by a small ε would still hold party e entering to the left of x_1. Thus, the payoff of party 2 will increase to $\pi_2(x_1, x_2 - \varepsilon) = 1 - F(\alpha - \frac{\varepsilon}{2}) > \pi_2(x)$, while the support of party 1 will drop to $F(\alpha - \frac{\varepsilon}{2}) - F(x_1) < \pi_1(x)$, a contradiction to the fact that x is an $\mathcal{R}$-equilibrium.

(ii). Suppose now that party e enters between x_1 and x_2, which will only happen if $z_1(x) > z_2(x)$. Then $D_1(x) = (z_1(x), z_2(x))$, and party e maximizes its support over this interval. Thus, if any of the established parties makes a "slight" move towards its competitor, party e would still enter "in between" and, by Lemma 4.2 in [10], the party initiating the move, will increase its support, a contradiction. Thus, $D_1(x)$ is empty.

(iii). Consider now the case, in which set $D_1(x)$ is empty, whereas set $D_{12}(x)$ consists of a unique element, x_1. Indeed, the entrant cannot share the first place by entering either to the left of x_1 or to the right of x_2. If $z_1(x) \le z_2(x)$, then there is no position between x_1 and x_2 that guarantees the entrant the share of the first place. If $z_1(x) > z_2(x)$, then, contrary to our assumption, set $D_1(x)$ is nonempty. Thus, the entrant

[11] When $z^\star$ is equal either to x_1 or to x_2, it represents the limit of "almost" vote-maximizing positions of the entrant.

selects her position x_e at x_1 and we have $\frac{1}{2}F(\alpha) > F(1-\alpha)$ and $x_1 = \frac{\alpha}{2}$. Then let party 2 leapfrog to the immediate left of x_1, thus, allowing the entrant to move to the immediate right of x_1. This move completely squeezes party 1, thus guaranteeing party 2 the sole possession of the second place, rather than being the last before the move took place. □

Proof of Proposition 3.1. Let a pair of strategies $x = (x_1, x_2)$ be an $\mathcal{R}$-equilibrium. In the view of Lemma 5.3, it remains to show that the set $D_2(x)$ is empty.

It suffices to demonstrate that $\mathbf{I}_1 = \mathbf{I}_2$ and $\mathbf{I}_3 = \mathbf{I}_4$ (Recall that $\mathbf{I}_1 \geq \mathbf{I}_4$.) Indeed, in this case, the same arguments as in the proof of Proposition 3.2 would yield the emptiness of the set $D_2(x)$.

Assume, in negation, that, at least, one of equalities $\mathbf{I}_1 = \mathbf{I}_2$ or $\mathbf{I}_3 = \mathbf{I}_4$ does not hold. There are several cases to consider.

(1) $\mathbf{I}_1 > \mathbf{I}_2$ and $\mathbf{I}_3 \leq \mathbf{I}_4$. Lemma 5.1 implies that the entrant could not displace one of the established parties by entering between x_1 and x_2. Depending on the relationship between $\mathbf{I}_1$, $\mathbf{I}_2$, $\mathbf{I}_3$ and $\mathbf{I}_4$, party e still may enter either to the left of x_1 or to the right of x_2.

(1a) $\mathbf{I}_1 > \mathbf{I}_4$. Party e would enter to the left of x_1, then x is not an $\mathcal{R}$-equilibrium by the same argument as in case (i) in Lemma 5.3.

(1b) $\mathbf{I}_1 = \mathbf{I}_4 > \mathbf{I}_3$. Party e would enter with equal probability to the left of x_1 and to the right of x_2, as both actions lead to being ranked as the second with equal electoral support. Hence

$$\pi_1(x) = \frac{F(\alpha) - F(x_1)}{2} + \frac{F(\alpha)}{2} = F(\alpha) - \frac{F(x_1)}{2},$$

recall, $\alpha = \frac{x_1+x_2}{2}$. Then there exists $\varepsilon > 0$ such that a "slight" shift of party 1 to its left, to $x_1 - \varepsilon$ would force party e to enter to the right of x_2, yielding $\pi_1(x_1 - \varepsilon, x_2) = F(\alpha - \frac{\varepsilon}{2}) > \pi_1(x)$, again a contradicting x being an $\mathcal{R}$-equilibrium.

(1c) $\mathbf{I}_1 = \mathbf{I}_4 = \mathbf{I}_3$. Party e would enter to the left of x_1, thus assuring the second place and with support of almost $\mathbf{I}_1$. In this case $\pi_1(x) = F(\alpha) - F(x_1)$. Then a move of party 1 to the left to x_1 leads to an entry of party e to the left of x_2, leaving the entrant in the second place with a higher support. Since the shift of party 1 can be chosen arbitrarily small, one can guarantee that party e enters "very close" to x_2. Thus, there exists $\varepsilon > 0$ such that $\pi_1(x_1 - \varepsilon, x_2) > \pi_1(x)$, a contradiction.

(2) $\mathbf{I}_1 \geq \mathbf{I}_2$, $\mathbf{I}_3 < \mathbf{I}_4$. Could be examined in the same manner as the case (1).

(3) $\mathbf{I}_1 < \mathbf{I}_2$, $\mathbf{I}_3 \leq \mathbf{I}_4$. Coupled with assumption (7) it implies $\mathbf{I}_2 > \mathbf{I}_4$. Party e would enter between x_1 and x_2. Indeed, by entering to the right of x_1, it could generate the support of "almost" $\mathbf{I}_2$ voters, thus assuring the second place. On the other hand, outside of this region it could attract the

support of no more than $\mathbf{I}_1 < \mathbf{I}_2$ voters, which gives her the second rank at best with less support. Moreover, since $\mathbf{I}_4 \geq \mathbf{I}_3$, Lemma 5.1 implies that the entrant is unable to attract more votes than party 2, by entering between x_1 and x_2, so that $D^m(x) = (x_1, z^1(x))$. Thus, the entrant should maximize the electoral support on that interval. The payoffs of both players in game Γ will be determined by

$$\pi_1(x) = \begin{cases} \mathbf{I}_1 & \text{if} \quad z^\star = x_1 \\ F(\frac{x_1+z^\star(x)}{2}) & \text{if} \quad z^\star(x) \in D^m(x) \\ F(\frac{x_1+z^1(x)}{2}) & \text{if} \quad z^\star(x) \notin D^m(x) \end{cases}$$

$$\pi_2(x) = \begin{cases} \mathbf{I}_3 + \mathbf{I}_4 & \text{if} \quad z^\star(x) = x_1 \\ 1 - F(\frac{x_2+z^\star(x)}{2}) & \text{if} \quad z^\star(x) \in D^m(x) \\ 1 - F(\frac{x_2+z^1(x)}{2}) & \text{if} \quad z^\star(x) \notin D^m(x). \end{cases}$$

Suppose now that party 2 moves its position to $\bar{x}_2$, which is "slightly" to the left of x_2. This would shrink the set of potential "in between" entry positions of party e, i.e., $D^m(x_1, \bar{x}_2) \subset D^m(x)$, as the aforementioned move of party 2 would shift y^1 to the left by assumption (A.2) [combining (13) and (10)], and therefore $z^1(x_1, \bar{x}_2) < z^1(x)$. Thus, if $z^\star(x) \geq z^1(x)$, this move yields a higher payoff to party 2. If $z^\star(x) < z^1(x)$, then Lemma 4.2 in [10] implies that party 2 benefits from its move, a contradiction.

(4) $\mathbf{I}_1 \geq \mathbf{I}_2, \mathbf{I}_3 > \mathbf{I}_4$. Could be examined in the same way as case (3) .
The last case to be considered is:

(5) $\mathbf{I}_1 < \mathbf{I}_2, \mathbf{I}_3 > \mathbf{I}_4$. Party e enters in the interval $[x_1, x_2]$, as $D(x) \subset [x_1, x_2]$ in this case. By using the previous arguments, we have
$D^m(x) = (x_1, z^1(x)) \bigcup (z^2(x), x_2)$, and, moreover, in the view of Lemma 5.3, we are left with the case $z^1(x) \leq z^2(x)$.
If $z^\star(x)$ is either less than $z_1(x)$ or greater than $z_2(x)$, then the consideration is the same as in the case (ii) of Lemma 5.3.
Assume, therefore, that $z_1(x) \leq z^\star(x) \leq z_2(x)$.

(5a) Let us first consider the case of $s^e(x; z_1(x)) \neq s^e(x; z_2(x))$ and, without loss of generality, $s^e(x; z_1(x)) > s^e(x; z_2(x))$. Thus, party e enters to the left of $z_1(x)$. Then, by moving to the left of x_2, party 2 would increase its support by forcing party e to shift its vote-maximizing position to the left by Lemma 4.2 in [10].

(5b) Let $s^e(x; z_1(x)) = s^e(x; z_2(x))$ and $z^\star(x) = z_1(x) = z_2(x)$, and so $s^e(x; z^*(x)) = 1/3$ and $D^m(x) = (x_1, z^\star(x)) \bigcup (z^\star(x), x_2)$. The best the entrant can do in this case is to displace one of the incumbents by entering as close as possible to $z^\star(x)$, and, being indifferent between entering slightly to the left of $z_1(x)$ or slightly to the right of $z_2(x)$, she chooses both actions with equal probability.

Let us show that x is not an $\mathcal{R}$-equilibrium. Take a "small" $\varepsilon > 0$ and consider two alternative moves:

(ζ): Party 1 moves to the right to $x_1' = x_1 + \varepsilon$. Denote $x_\zeta = (x_1', x_2) \in \mathbb{R}^2$.

(η): Party 2 moves to the left to $x_2' = x_2 - \varepsilon$. Denote $x_\eta = (x_1, x_2') \in \mathbb{R}^2$.

First consider ζ. By assumption (A.2), there are $z_\zeta^1 = z^1(x_\zeta) < z^\star(x) < z_\zeta^2 = z^2(x_\zeta)$, such that the set of possible entry positions of party e consists of two disjoint intervals: $D^m(x_\zeta) = (x_1, z_\zeta^1) \bigcup (z_\zeta^2, x_2) \subset \mathbb{R}$. In this case, again, only one of the established parties can be replaced, so provided party e can only guarantee to be the second, it maximizes voters' support. The best position of party e under ζ, $z_\zeta^\star = z^\star(x_\zeta)$, should by to the right of $z^\star(x)$ by Lemma 4.2 in [10]. Therefore, $z_\zeta^\star > z_\zeta^1$. If $s^e(x_\zeta; z_\zeta^1) < s^e(x_\zeta; z_\zeta^2)$, the entrant will replace party 2, thus benefiting party 1, a contradiction.

It is left to consider the case, in which $s^e(x_\zeta; z_\zeta^1) \geq s^e(x_\zeta; z_\zeta^2)$ after the move of the first party. Note that the last inequality could be rewritten as

$$s^e(x_\zeta; z_\zeta^1) = F\left(y^1(r - \frac{\varepsilon}{2})\right) \geq 1 - F\left(y^2\left(r - \frac{\varepsilon}{2}\right)\right) = s^e(x_\zeta; z_\zeta^2). \tag{15}$$

The examination of the move η is similar. Again assumption (A.2) yields the existence of $z_\eta^1 = z^1(x_1, x_2)$ and $z_\eta^2 = z^2(x_\eta)$, such that $D^m(x_\eta) = (x_1, z_\eta^1) \bigcup (z_\eta^2, x_2)$. Again we assume that $z_\eta^\star < z_\eta^2$ and $s^e(x_\eta; z_\eta^2) \geq s^e(x_\eta, z_\eta^1)$. Thus, we have

$$s^e(x_\eta; z_\eta^2) = F\left(y^1(r - \frac{\varepsilon}{2})\right) \geq 1 - F\left(y^2\left(r - \frac{\varepsilon}{2}\right)\right) = s^e(x_\eta; z_\eta^1). \tag{16}$$

Combining (15) and (16), we conclude that party e generates the support of the same number of voters, denoted by s_0, whenever it chooses one of the following positions z_ζ^1 or z_ζ^2 under ζ and z_η^1 or z_η^2 under η. Note also that since the functions $y^1(\cdot)$ and $y^2(\cdot)$ depend *only* on the distance between the positions of the first two parties, the set of voters who pick z_ζ^1 (z_ζ^2, respectively) under ζ is the same as of those whose best choice z_η^1 (z_η^2, respectively) under η. Hence party 1 would be better off under ζ than under η, whereas the opposite is true for party 2, i.e.,

$$\pi_1(x_\zeta) > \pi_1(x_\eta); \;\; \pi_2(x_\eta) > \pi_2(x_\zeta). \tag{17}$$

As both moves generate support of s_0 voters for party e, we also have

$$\pi_1(x_\zeta) + \pi_2(x_\zeta) = \pi_1(x_\eta) + \pi_2(x_\eta) = 1 - s_0. \tag{18}$$

Combining (17) and (18) we obtain $\pi_1(x_\zeta)+\pi_2(x_\eta) > 1-s_0$. However, by Assumption (A.2), both moves ζ and η lead to a decline in support of party e, i.e., $s_0 < \frac{1}{3} = s^e(x; z^*(x))$. Thus,

$$\pi_1(x_\zeta) + \pi_2(x_\eta) > \frac{2}{3}$$

It follows, therefore, that at least one of the numbers $\pi_1(x_\zeta)$ or $\pi_2(x_\eta)$ exceeds $\frac{1}{3}$, yielding either $\pi_1(x_\zeta) > \pi_1(x)$ or $\pi_2(x_\eta) > \pi_2(x)$. That is, at least one of the established parties would benefit by deviating, a contradiction.

(5c) Finally, let $s^e(x; z_1(x)) = s^e(x; z_2(x))$ with $z_1(x) < z_2(x)$, in which case party e enters with equal probability to the left of $z_1(x)$ and to the right of $z_2(x)$. But then the similar arguments as in the consideration of the previous case show that x is not an $\mathcal{R}$-equilibrium. □

Proof of Proposition 3.3. By Proposition 3.2, there is a unique pair of $\mathcal{D}$-strategies (x_1^d, x_2^d). Then, by (2) and (3), $F(x_1^d) = \frac{1}{2}F(\frac{x_1^d+x_2^d}{2})$ and $1 - F(x_2^d) = \frac{1}{2}(1 - F(\frac{x_1^d+x_2^d}{2}))$. Assume that (4) holds, i.e., $F(\frac{x_1^d+x_2^d}{2}) = \frac{1}{2}$. Then the payoff of each player in game Γ is equal to $\frac{1}{2}$. We shall show that if one of the players chooses a different strategy, her payoff will decline.

Consider party 1. Suppose first that it moves to the right of x_1^d by choosing $x_1 > x_1^d$. Since this move of party 1 shrinks the mass of voters located between the incumbents' positions, party e could not displace an established party by entering between x_1 and x_2^d. Hence, the entrant should come either to the left of x_1 or to the right of x_2^d. Since $F(x_1) > \frac{1}{4} = 1 - F(x_2^d)$, the optimal entrant's move would be to the left and "very close" to x_1. Then the payoff of player 1 in game Γ would be less than $F(x_2^d) - F(x_1) < \frac{1}{2}$. Thus, player 1 would be worse off by moving to the right of x_1^d.

Suppose now that party 1 moves to the left of x_1^d by choosing $x_1 < x_1^d$. Since the deviation of party 1 expands the mass of voters located between the incumbents' positions, party e could not displace an established party by entering to the right of x_2^d. Moreover, by assumption (A.2) (and by the argument analogous to that in Lemma 5.1), $F(\frac{x_1+x_2^d}{2}) - F(x_1) > \frac{1}{4}$ so that party e could not displace party one, and, clearly, not the second party whose support is over $1/2$. Therefore, an entrant will not choose a platform to the left of x_1, either. Since, by Proposition 3.1, $D(x_1, x_2^d)$ is nonempty, the entrant could displace one of the established parties by entering between x_1 and x_2^d. Then the payoff of player 1 would be less than $F(\frac{x_1+x_2^d}{2}) < \frac{1}{2}$ making her worse off. This completes the "if" part of the Proposition.

Suppose that (4) does not hold and assume, without loss of generality, that $F(x_1^d) < 1 - F(x_2^d)$. Choose $\delta > 0$ such that $F(x_1^d + \delta) < 1 - F(x_2^d)$ and consider the move of party 1 to $x_1 = x_1^d + \delta$. By entering between x_1 and x_2^d, as well as by entering to the left of x_1 the entrant would get less than a quarter of the votes. However, by entering to the right and "very close" to x_2^d, party e would displace party 2 and would receive more than 1/4 of the votes. Moreover, the support of the first party will increase, $\pi_1(x_1^d + \delta, x_2^d) = F(\frac{x_1^d+\delta+x_2^d}{2}) > F(\frac{x_1^d+x_2^d}{2}) = \pi_1(x_1^d, x_2^d)$. Thus, the party which gets less than 50% of the total vote in $\mathcal{D}$-strategies would be better off by moving towards its rival incumbent and allowing entry of the third party.

To conclude the proof of the proposition, note that by (2), (3), (4), the pair (x_1, x_2) constitutes an $\mathcal{R}$-equilibrium if and only if $F(x_1) = \frac{1}{4}$, $F(x_2) = \frac{3}{4}$ and $F(\frac{x_1+x_2}{2}) = \frac{1}{2}$. □

References

1. Alvarez, R.M., Nagler, J.: A new approach for modeling strategic voting in multi-party elections. Br. J. Pol. Sci. **30**, 57–75 (2000)
2. Bergstrom, T., Bagnoli, M.: Log-concave probability and its applications. Econ. Theory **26**, 445–469 (2005)
3. Cohen, R.: Symmetric 2-equilibria of unimodal voter distribution curves. Mimeo, Harvard University 1985
4. Eaton, B., Lipsey, R.: The principle of minimum product differentiation reconsidered: Some new developments in the theory of spatial competition. Rev. Econ. Stud. **42**, 27–49 (1975)
5. Gale, D., Nikaido, H.: The Jacobian matrix and global univalence of mappings. Math. Ann. **159**, 81–93 (1965)
6. Green, D., Palmquist, B., Schickler, E.: Partisan Hearts and Minds: Political Parties and the Social Identities of Voters. Yale University Press, New Haven 2002
7. Greenberg, J., Shepsle, K.: The effect of electoral rewards in multiparty competition with entry. Am. Pol. Sci. Rev. **81**, 525–537 (1987)
8. Palfrey, T.: Spatial equilibrium with entry. Rev. Econ. Stud. **51**, 139–156 (1984)
9. Parthasarathy, T.: On Global Univalence Theorems. Lecture Notes in Mathematics. Springer, Berlin Heidelberg New York 1983
10. Weber, S.: On hierarchical spatial competition. Rev. Econ. Stud. **59**, 407–425 (1992)
11. Weber, S.: Entry deterrence in electoral spatial competition. Soc. Choice Welfare **15**, 31–56 (1997)

Adv. Math. Econ. 10, 121–122 (2007)

Advances in
MATHEMATICAL
ECONOMICS

Publisher's Errata

Solving long term optimal investment problems with Cox-Ingersoll-Ross interest rates*

H. Hata and J. Sekine

[1] Graduate School of Engineering Science, Osaka University, Toyonaka, Osaka 560-8531, Japan
(e-mail: hata@sigmath.es.osaka-u.ac.jp)
[2] Institute of Economic Research, Kyoto University, Yoshida-Honmachi, Sakyo-ku, Kyoto 606-8501, Japan
(e-mail: sekine@kier.kyoto-u.ac.jp)

The following sentences were printed incorrectly in the above-mentioned article due to the publisher's error.

1. On p. 240, on line 2 from the bottom,

The effective domaineffective domain

should be:

The effective domain

* The original article was published in the *Advances in Mathematical Economics* volume 8, p.231–p.255. Its online version can be found at http://www.hiranolab.jks.ynu.ac.jp/ advances/4/current.html

2. On p. 241, on line 2,

The difference $\triangle_T$ of growth rategrowth rates
should be:
The difference $\triangle_T$ of growth rates

On the same page, on the bottom line,

given by (1.3) is nearly optimalnearly optimal
should be:
given by (1.3) is nearly optimal

3. On p.246, on line 14 from the bottom,

So from Feynman–Kac's formulaFeynman–Kac's formula
should be:
So from Feynman–Kac's formula

Subject Index

Instructions for Authors

A. General

1. Papers submitted for publication will be considered only if they have not been and will not be published elsewhere without permission from the publisher and the Research Center for Mathematical Economics.

2. Every submitted paper will be subject to review. The names of reviewers will not be disclosed to the authors or to anybody not involved in the editorial process.

3. The authors are asked to transfer the copyright to their articles to Springer if and when these are accepted for publication.

The copyright covers the exclusive and unlimited rights to reproduce and distribute the article in any form of reproduction. It also covers translation rights for all languages and countries.

4. Manuscript must be written in English. One original and 3 sets of photocopies should be submitted to the following address.

Professor Toru Maruyama
Department of Economics
Keio University
2-15-45 Mita, Minato-ku, Tokyo
108-8345 Japan

5. **Offprints**: Twenty-five (25) **offprints** of each paper will be supplied free of charge. Additional offprints can be ordered at cost price.

B. Preparation of Manuscript

1. Manuscripts must be double-spaced (not 1.5), with wide margins (at least 25 mm), and large type (at least 12 point) on one side of A4 paper. Any manuscript that does not meet these requirements will be returned to the author immediately for retyping.

2. All manuscripts would finally be composed using our Springer LaTeX macro package. The Springer macros use standard LaTeX packages - if you do not have these packages refer to `http://www.ctan.org` or `http://www.dante.de/tex`. If authors make manuscripts by word-processing software other than TeX, please follow our styles as possible. For authors who prepare their manuscripts by TeX, we strongly recommend to visit our homepage, `http://www.springer.com/economics/`. You can download all the necessary macro packages with instructions for LaTeX2e from `ftp://ftp.springer.de/pub/tex/latex/svjour/global/`. For support, please contact `texhelp@springer.de`.

3. **The title page** must include: title; short (running) title of up to 60/85 characters; first names and surnames of all coauthors with superscript numerals indicating their affiliations; full street addresses of all affiliations; address to which proofs should be sent; fax number; and any footnotes referring to the title (indicated by asterisks*).

4. **Summary/abstract**: Each paper must be preceded by a summary/an abstract, which should not exceed 100 words.

5. **The Journal of Economic Literature index number (JEL classification)** should be indicated and the statement of the **2000 Mathematics Subject Classification (MSC) numbers** is desirable. You can check JEL classification with Internet at `http://ideas.repec.org/JEL/` as well as 2000 MSC numbers at `http://www.ams.org/msc`.

6. **Main text**: All tables and figures must be cited in the text and numbered consecutively with Arabic numerals according to the sequence in

which they are cited. Please mark the desired position of tables and figures in the left margin of the manuscript.

Do not italicize, underscore, or use boldface for headings and subheadings. Words that are to be italicized should be underscored in pencil.

Abbreviations must be spelled out at first mention in summary/abstract and main text. Abbreviations should also be spelled out at first mention in figure legends and table footnotes.

Short equations can be run in with the text. Equations that are displayed on a separate line should be numbered.

7. **References**: The list of references should be in alphabetical order and include the names and initials of all authors (see examples below). Whenever possible, please update all references to papers accepted for publication, preprints or technical reports, giving the exact name of the journal, as well as the volume, first and last page numbers and year, if the article has already been published or accepted for publication.

When styling the references, the following examples should be observed:

Journal article:
1. or [F-M] Freed, D.S., Melrose, R.B.: A mod k index theorem. Invent. Math. **107**, 283-299 (1992)

Complete book:
2. or [C-S] Conway, J.H., Sloane, N.J.: Sphere packings, lattices, and groups (Grundlehren Math. Wiss. Bd. 290) Berlin Heidelberg New York: Springer 1988

Single contribution in a book:
3. or [B] Border, K.C.: Functional analytic tools for expected utility theory. In: Aliprantis, C.D. et al. (eds.): Positive operators, Riesz spaces and economics. Berlin Heidelberg New York: Springer 1991, pp. 69-88

8. **Citations in the text** should be either by numbers in square brackets, e.g. [1], referring to an alphabetically ordered and numbered list, or by the author's initials in square brackets, e.g. [F-M], or by author and year in parentheses, e.g. Freed and Melrose (1992). Any of these styles is acceptable if used consistently throughout the paper. In the third system, if a work with more than two authors is cited, only the first author's name plus "et al." need be given and if there is more than one reference by the same author or team of authors in the same year, then a, b, c should be added after the year both in the text and in the list of references.

9. **Tables** are to be numbered separately from the illustrations. Each table should have a brief and informative title. All abbreviations used in a table must be defined in a table footnote on first use, even if already defined in the text. In subsequent tables abbreviations need not be redefined. Individual footnotes should be indicated by superscript lowercase a, b, c, etc. Permission forms must be provided for any tables from previously published sources (same procedure as with figures; see below).

10. **Figures**: If you have access to suitable software, you can design your figures directly on a computer, but creation by other means is of course also possible. In any case the originals should be pasted into the camera-ready copy at the appropriate places and with the correct orientation. If necessary, the figures may be drawn overscale, but in this case suitably reduced copies should be pasted into the script.

If a figure has been published previously, acknowledge its source and submit written permission signed by author and publisher. The source must be included in the reference list. If a permission fee is required, it must be paid by the author. Responsibility for meeting this requirement lies entirely with the author.

MIX
Papier aus verantwortungsvollen Quellen
Paper from responsible sources
FSC® C105338

If you have any concerns about our products,
you can contact us on
ProductSafety@springernature.com

In case Publisher is established outside the EU,
the EU authorized representative is:
Springer Nature Customer Service Center GmbH
Europaplatz 3, 69115 Heidelberg, Germany

Printed by Libri Plureos GmbH
in Hamburg, Germany

The Large Dam Dilemma

Pu Wang • Shikui Dong • James P. Lassoie

The Large Dam Dilemma

An Exploration of the Impacts of Hydro Projects on People and the Environment in China

Springer

Pu Wang
Department of Natural Resources
Cornell University
Ithaca, NY 14853
USA

Shikui Dong
School of Environmental Sciences
Beijing Normal University
Beijing, 100875
China

James P. Lassoie
Department of Natural Resources
Cornell University
Ithaca, NY 14853
USA

ISBN 978-94-007-7629-6 ISBN 978-94-007-7630-2 (eBook)
DOI 10.1007/978-94-007-7630-2
Springer Dordrecht Heidelberg New York London

Library of Congress Control Number: 2013949436

Printed on acid-free paper

Springer is part of Springer Science+Business Media (www.springer.com)

Preface

Published in 1987, the Brundtland Report, Our Common Future, culminated over 2 decades of concern and debate about environmental degradation and human welfare and posed a challenge for societies worldwide to seek sustainable approaches to development. Solutions to the Earth's most difficult and recalcitrant environmental problems were identified as falling at the nexus of the scientific disciplines, which set the agenda for developing interdisciplinary approaches to research and development that continues today. The relatively simplistic notion of "interdisciplinarity" promoting the optimization of environmental, social, and economic needs has now been replaced by the more complicated and realistic concept of "coupled human and natural systems (CHANS)." The CHANS theory approach interconnects human activities and ecosystem functioning and acknowledges the true complexities facing environmental conservation and sustainable development in the twenty-first century.

Meeting the needs of a growing human population for economic development without compromising the long-term integrity of the environmental foundation for all life is the essence of society's search for sustainable development strategies. Among the many environmental challenges facing humankind, including global climate change, deforestation, rangeland degradation, energy production, and loss of biodiversity, those related to the globe's water resources are perhaps the most acute. People sicken and perish and nations stagnate and decline without adequate supplies of clean freshwater. Many believe that water may have already replaced oil as the Earth's most precious and endangered liquid and that the formation of international "water cartels" to control its distribution is not far off.

Water and energy come together in the building of large dams intended for the generation of hydroelectricity. Often termed "clean energy" large-scale hydropower projects are not without their sustainability critics. While it is difficult to counter their ability to produce efficient and effective power in the service of national economic development, the construction of dams and power stations are also known for their negative environmental impacts and mixed effects on the socioeconomic conditions of local people. Hence, it is not surprising that proposals promoting the construction of major hydroelectric dams often face public and scientific scrutiny that sometimes leads to widespread civil unrest, protests, and legal and political repercussions. Understanding the social, economic, and environmental impacts arising from large-scale hydro-projects is a much-needed step toward sustainable development, not as a deterrent to their construction, but as a means for reducing their potentially negative effects.

I am very pleased to see the publication of this timely and informative book, The Large Dam Dilemma: An Exploration of the Impacts of Hydro Projects on People and the Environment in China. I congratulate the authors for their presentation of new research findings from a recent study mainly on the Upper-Mekong River, and their synthesis of other investigations of large dams elsewhere in China. China is currently leading the world in the construction of large-scale dams and it is most appropriate that China also takes the lead in helping improve the sustainability of such important hydro-projects. This book provides a compendium of information and insights that will prove valuable during the planning phase for such projects worldwide. I highly recommend it to scientists, planners, government officials, and public organizations concerned about the protection and sustainable development of the Earth's fragile water resources.

Beijing, People's Republic of China Hao Wang

Academician, Chinese Academy of Engineering
Professor, China Institute of Water Resources
and Hydropower Research

Contents

Abstract

Large dam construction has significant environmental and social impacts at different scales. This book first summarizes and updates information about the history, distribution, functions, and impacts of large dams, both globally and at China's national level. It then addresses the environmental and social impacts of large dams in China and introduces an empirical study conducted during 2010 in areas affected by dams along the Upper-Mekong River, China. We present innovative methods for assessing the impacts of dams on biological diversity at the watershed scale and impacts on ecological integrity of rivers at the ecosystem scale. Then we developed a framework to assess the impacts of dam construction on different dimensions of wealth of dam-affected households, namely material (land, houses), embodied (knowledge, skills), and relational (infrastructure) wealth; and compare losses and compensations for each dimension. Results indicated that large dam construction has significant negative impacts on biological diversity and ecological integrity; local communities often suffer from wealth loss in all three characterized dimensions, but government compensation policies typically consider only material wealth; and this inequity leads to dissatisfaction on both sides and is the root cause for disagreements and conflicts. These results will prove important to future dam projects in China, and possibly elsewhere, as they suggest that more comprehensive environmental and social impact assessments are needed for large dam projects, and less dissatisfaction will arise from community relocation projects when the affected villagers and decision-makers acknowledge and agree on the degree of losses and resulting compensations in all three dimensions of wealth.

Keywords Large dam • environmental impact • social impact • biodiversity • resettlement • compensation policy

Introduction: The Large Dam Dilemma

Food, water, and energy, three of the most critical issues for human development, are all connected with one facility—large dams. By International Commission on Large Dams' (ICOLD) definition, large dams are those with heights over 15 m (WCD 2000). It is estimated that 30–40 % of irrigated land around the world relies on dams, and irrigated land contributes approximately 40 % of the world agricultural production (WCD 2000; Shah and Kumar 2008). Large dams also have guaranteed water security in many urban and industrialized regions, with reports showing high positive relationships between dam density and water security level (Vorosmarty et al. 2010). They have also been used as an important way to control floods, and presently about 13 % of existing dams have flood control functions (WCD 2000). Electricity generation is another important reason for building large dams, and about 19 % of electricity worldwide is generated by hydropower dams; in 63 countries, hydropower supplies more than 50 % of the electricity (WCD 2000).

Even though large dams have been used as a means of development for a long time, they also have caused various environmental and social problems at different scales. Dams block water and alter natural flow regimes of rivers, which has significant impacts on river ecosystems and fisheries (Poff et al. 1997); the reservoirs formed after dam construction submerge farmland and terrestrial ecosystems (WCD 2000); and dams change the geological conditions of reservoir areas, having the potential to trigger landslides and earthquakes (Kerr and Stone 2009; Pandey and Chadha 2003; Deji 1999). In addition, the negative social consequences of large dams include the effects of millions of people being relocated or displaced, uneven benefit and cost distribution among different groups, and impacts on indigenous and tribal people and their cultures (Égré and Senécal 2003; Tilt et al. 2009).

After the rise of environmentalism (1960s), and especially after the concept of sustainable development became prevalent (1980s), the benefits and costs of building dams are now more comprehensively scrutinized. People first began to question the rationale for using large dams to promote development in the 1970s, and the dam debate intensified worldwide during the 1990s. Since, the rate of dam construction has slowed markedly in developed countries, and in the United States, the rate of decommissioning old dams actually exceeded the rate of construction by 1998 (WCD 2000). However, in developing countries, there are still many large dams under construction or being planned.

Chinese experience in dam construction is largely consistent with the rest of the world (Wang et al. 2013a). China's large dam construction began in the second half of twentieth century, when modernization and developmentalism became the dominant ideologies globally. Large dams became a symbol of reconstruction and development and were used as means for river basin development, flood control, and electricity generation. But because of China's socioeconomic and political peculiarities, Chinese dam construction has its own unique characteristics. As the most populous nation and the second largest and fast growing economy in the world, China has built more large dams than any other country. While the anti-dam movement is increasing in developed countries, China is still ambitiously developing hydropower as part of its long-term national plan. The environmental and social impacts of dams and the involvements of markets and civil society in the decision-making processes for dam construction in China also have different features. Hence, research on Chinese dam issues has both national and global significance.

The future of large dams in the world and in China remains unclear, with many pertinent questions. For example, do we still need to rely heavily on dam-irrigated agriculture, or can applying water-saving technologies and using drought-resistant agricultural crops avoid global food crises? Should we continue to control floods with hydrologic engineering methods, which change the natural features of rivers, or do we appeal to more fundamental approaches, such as restoring degraded environments and reestablishing natural flow regimes? Is hydropower a renewable and clean alternative for fossil fuels, or does energy from dammed rivers have even higher environmental and social costs?

Comprehensive answers to these major questions are beyond the scope of this book, but we hope that our interdisciplinary assessment of large dams in China and elsewhere will identify pathways for doing so. Our approach is to review and synthesize environmental and socioeconomic information about large dams at different scales, including our own recent research on the Upper-Mekong River (Lancang River) in China (Wang et al. 2013a, b; Li et al. 2012a, c, 2013; Zhao et al. 2012a, b, c). We begin by providing a global perspective on large dams, including arguments from both sides of the ongoing debate over their construction (Chap. 1). We then turn our attention to large dams in China by first providing an overview and case studies (Chap. 2) followed by detailed assessments of their environmental (Chap. 3) and socioeconomic (Chap. 4) impacts. Having identified key points to consider when planning the construction of large dams, we conclude with a discussion of how the large dam debate is evolving into decision-making policies in China (Chap. 5). In sum, we hope that this book is a modest, but useful step toward identifying final solutions to the large dam dilemma worldwide.

References

Deji C (1999) Engineering geological problems in the Three Gorges Project on the Yangtze, China. Eng Geol 51:183–193

Égré D, Senécal P (2003) Social impact assessments of large dams throughout the world: lessons learned over two decades. Impact Assess Project Appr 21:215–224

Kerr RA, Stone R (2009) A human trigger for the great quake of Sichuan? Science 323:322

Li J, Dong S, Peng M, Li X, Liu S (2012a) Vegetation distribution pattern in the dam areas along middle-low reach of Lancang-Mekong River in Yunnan Province, China. Front Earth Sci 6:283–290

Li J, Dong S, Yang Z, Peng M, Liu S, Li X (2012c) Effects of cascade hydropower dams on the structure and distribution of riparian and upland vegetation along the middle-lower Lancang-Mekong River. For Ecol Manage 284:251–259

Li J, Dong S, Peng M, Yang Z, Liu S, Li X, Zhao C (2013) Effects of damming on the biological integrity of fish assemblages in the middle Lancang-Mekong River basin. Ecol Indicators 34:94–102

Pandey AP, Chadha RK (2003) Surface loading and triggered earthquakes in the Koyna-Warna region, western India. Phys Earth Planet Int 139:207–223

Poff NL, Allan JD, Bain MB, Karr JR, Prestegaard KL, Richter BD, Sparks RE, Stromberg JC (1997) The natural flow regime. Bioscience 47:769–784

Shah Z, Kumar M (2008) In the midst of the large dam controversy: objectives, criteria for assessing large water storages in the developing world. Water Res Manage 22:1799–1824

Tilt B, Braun Y, He D (2009) Social impacts of large dam projects: a comparison of international case studies and implications for best practice. J Environ Manage 90:S249–S257

Vorosmarty CJ, McIntyre PB, Gessner MO, Dudgeon D, Prusevich A, Green P, Glidden S, Bunn SE, Sullivan CA, Liermann CR, Davies PM (2010) Global threats to human water security and river biodiversity (vol 467, pg 555, 2010). Nature 468:334

Wang P, Lassoie JP, Dong S, Morreale SJ (2013a) A framework for social impact analysis of large dams: a case study of cascading dams on the Upper-Mekong River, China. J Environ Manage 117:131–140

Wang P, Wolf SA, Lassoie JP, Dong S (2013b) Compensation policy for displacement caused by dam construction in China: an institutional analysis. Geoforum 48:1–9

WCD (2000) Dams and development: a new framework for decision-making. World Commission on Dams

Zhao Q, Liu S, Deng L, Dong S, Yang J, Wang C (2012a) The effects of dam construction and precipitation variability on hydrologic alteration in the Lancang River Basin of southwest China. Stoch Environ Res Risk Assess 26:993–1011

Zhao Q, Liu S, Deng L, Dong S, Yang Z, Yang J (2012b) Landscape change and hydrologic alteration associated with dam construction. Int J Appl Earth Obs Geoinf 16:17–26

Zhao Q, Liu S, Deng L, Yang Z, Dong S, Wang C, Zhang Z (2012c) Spatio-temporal variation of heavy metals in fresh water after dam construction: a case study of the Manwan Reservoir, Lancang River. Environ Monit Assess 184:4253–4266

Chapter 1
A Global Review of Large Dam Construction

1.1 History and Distribution of Large Dams in the World

The history of dam construction is as old as human civilization. In ancient Chinese mythology, the legendary tribal leader, Yao, assigned one of his subordinates, Gun, to lead the people to fight a great flood. Gun built dams to block the flood, but they broke and caused an even larger disaster. As punishment, Yao's successor, Shun, killed Gun, and then assigned Gun's son, Yu, to lead the fight against the flood. Yu removed the dams and other barriers blocking the river, allowing the water to flow freely to the ocean, and finally overcame the flood. This story shows the long history of the impacts of dam construction on attempts to control the flow of rivers.

The remains of dams dating back to 6000 BP have been found in Mesopotamia; and irrigation and water supply dams became widespread in many parts of the world by 2000 BP (WCD 2000). The oldest continuously functioning dam is likely one associated with the Dujiangyan Irrigation Project in Sichuan Province, China, which was built in 256 BP and is still providing irrigation water for a large area of farmland on the Chengdu Plain (Zhang and Jin 2008). The first hydropower dam in the world was built about 1890 in the United States (WCD 2000).

Before 1900, there were only about 700 large dams worldwide (Fig. 1.1). Most of the large dams currently in existence were built during the twentieth century. About 5,000 large dams were built during the first half of the century, three-quarters of them in developed countries. Dam construction rapidly increased globally after the Second World War, with the peak occurring between 1970 and 1975, when nearly 5,000 large dams were built worldwide. The rate declined after the 1980s in most parts of the world, especially in North America and Europe (Shah and Kumar 2008; WCD 2000).

According to the World Bank, countries represented by the Organization for Economic Co-operation and Development (OECD) have already developed 70 % of their economically feasible dam potentials, while developing countries are only using 30 % of their potentials. Strikingly, less than 10 % of the potential has been exploited in Africa (World Bank 2009). Since developed countries have built dams on most of

P. Wang et al., *The Large Dam Dilemma: An Exploration of the Impacts of Hydro Projects on People and the Environment in China*, DOI 10.1007/978-94-007-7630-2_1,

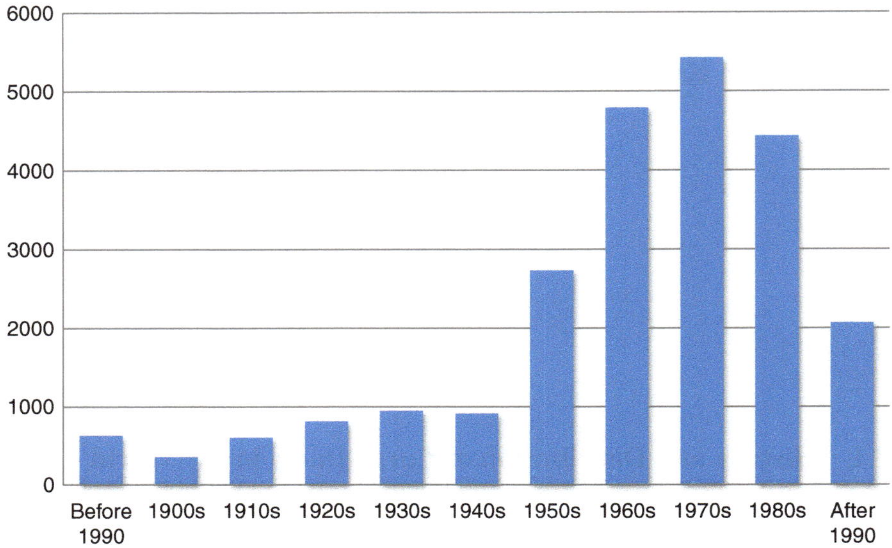

Fig. 1.1 Construction of dams by decade (Data from WCD (2000); *Source*: ICOLD, 1998, excluding over 90 % of large dams in China)

their rivers, they now mainly focus on managing and improving the efficiency of existing dams. But for developing countries, dam construction is considered a promising development opportunity. Therefore, the future of the large dam construction worldwide largely depends on the rate of development activities in these countries.

By the end of twentieth century, more than 45,000 large dams had been built around the world (WCD 2000). Fig. 1.2 shows the worldwide distribution of large dams. Five regions, East Asia, South Asia, North America, Europe, and Southern Africa, have the highest density of large dams.

Asia has not only the most existing dams compared to all other regions, but is also currently experiencing the highest construction rates. China and India, the two most populous countries in the world, are building most of these new large dams. Rapid economic development in these two countries not only demands more energy and water, but also provides the financial resources for major construction projects, such as the building of large dams (Bawa et al. 2010; Liu and Diamond 2005). The top five countries with most of the dams currently under construction are all in Asia: India, with 700–900 new dams under construction; China, with 280; Turkey, with 209; South Korea, with 132; and Japan, with 90 (WCD 2000).

The United States has built the greatest number of large dams in North America, about 8,000. However, by the end of the twentieth century, a new trend was evident as old and poorly functioning dams were decommissioned across the United States. About 15 years ago, the top official at the US Bureau of Reclamation, which had been responsible for building massive dams throughout the American West, declared that the "era of big dams is over" (Longman 2008). In 1998, the speed of decommissioning exceeded that of construction for the first time (WCD 2000).

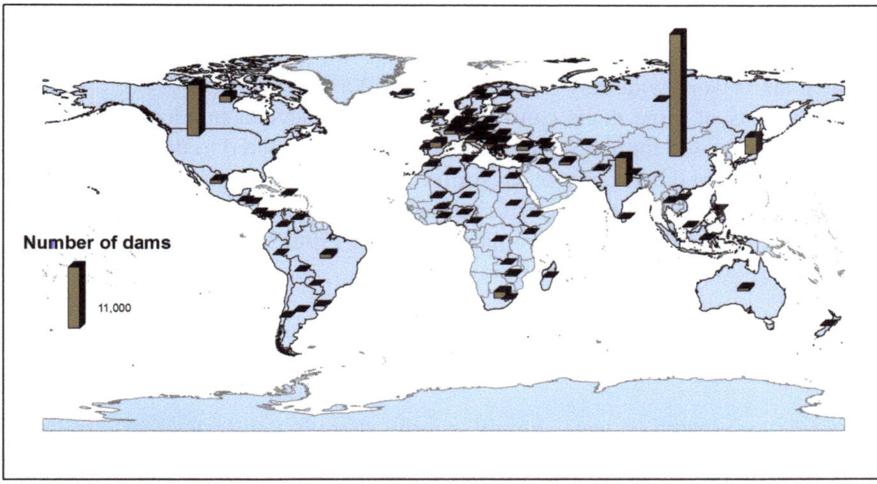

Fig. 1.2 Distribution of large dams in the world (Synthesized from data on ICOLD website: http://www.icold-cigb.org/)

Decommissioning old dams has many benefits for the restoration of riparian systems. However, some studies have shown that the removal of old dams may also cause negative impacts on associated ecosystems as they are forced to readapt to new hydraulic regimes (Poff and Hart 2002; Bednarek 2001) .

Currently, there are about 6,000 large dams in Europe. The pattern of construction of dams and reservoirs in Europe can be illustrated using the United Kingdom and Spain as examples. In the UK, the number of large dams grew rapidly during the nineteenth century, from fewer than 10–175, a rate of 1.7 per year. By 1950, the rate had almost doubled. After 1950, construction took place at a rate of 5.4 dams per year before dropping to zero by the late 1990s. Today, the UK has a total of 486 dams. In contrast, Spain saw the number of reservoirs grow at the rate exceeding 4 per year between 1900 and 1950, before almost doubling and reaching 741 facilities by 1975. By 1990, this figure had more than doubled again (19.5 per year), and today there are 1,172 large dams in Spain, the most in Europe (European Environmental Agency 2010). With suitable sites becoming fewer and environmental concerns becoming greater, the total number of dams in Europe is now growing very slowly.

1.2 Multiple Functions of Large Dams

1.2.1 Irrigation

Large dams have made significant contributions to rapid increases in yields associated with modern agriculture. About one fifth of the world's agricultural land is irrigated, and irrigated agriculture accounts for about 40 % of the agricultural production globally. Half of the world's large dams were built exclusively or primarily for irrigation, and an estimated 30–40 % of the 268 million hectares of irrigated lands worldwide rely on dams (WCD 2000).

The heights of large dams can range from 15 to several hundred meters. But generally, dams used for irrigation are relatively low compared to those constructed for flood control and hydropower. In North China, where precipitation is often inadequate for irrigation and the flow of rivers is generally low, many dams have been built to meet the water needs of cropland (Nickum 1998).

1.2.2 Water Supply

According to research on water supply security, there is a clear positive correlation between the density of dams and security level (Vorosmarty et al. 2010). Many reservoirs were built to provide a reliable supply of water to meet rapidly growing urban and industrial needs, especially in drought-prone regions where natural groundwater sources and existing lakes or rivers were considered inadequate to meet all needs. Globally, about 12 % of all large dams are designated as water supply dams and about 60 % of these are in North America and Europe (WCD 2000). Most metropolitan areas must rely heavily on reservoirs in distant surrounding watersheds for water supply, such as Beijing (Government of Beijing 1981) and New York City (The City of New York 2013).

1.2.3 Electricity Generation

Hydropower provides about 19 % of the world's electricity production, and in some countries, it is the most important power source (WCD 2000). Therefore, dams for hydropower are constructed globally as an important development approach. In addition to the ambitions of many different governments, the World Bank and other development banks also promote large hydropower dams by providing a large number of loans, especially to developing countries (World Bank 2009).

Hydropower is considered to be cleaner than electricity generated by burning fossil fuels. It has almost zero CO_2 emission (if the CO_2 emitted by deteriorating organic matter at the bottom of the reservoirs is not considered), and once completed, no additional inputs are needed other than the maintenance of the power station. In the view of hydraulic engineers, some deep valleys with rapid elevation drops are particularly suitable for building large dams.

The electricity generated from hydropower dams is most often fed into national grids, which benefits people in the whole country. However, local people commonly bear most of the negative impacts from dam construction and the operation of the power station (Magee 2006). This uneven cost-benefit distribution is one of the main equity issues related to large dams.

1.2.4 Flood Control

Worldwide, flooding has caused more deaths than any other natural disasters (WCD 2000). In those countries with monsoonal climates, flood control during the rainy season becomes even more imperative. Generally, flood control projects include not only dams and reservoirs, but also levees, all of which significantly change natural flow regimes.

However, dams constructed for flood control can sometimes make the situation even worse. For some rivers with high sand concentrations, like the Yellow River in China, dams often cause serious sedimentation problems, which lift the riverbed and reduce the water-holding capacity of the original watercourses (Xu 1998; Yang et al. 2008). The case study part in the second chapter will show the failure of Sanmenxia Dam on the Yellow River, which is a good illustration of how a flood control dam can exacerbate flooding disasters.

1.2.5 Navigation

Only a small number of large dams have been constructed specifically for the purpose of navigation. But for some, the need to improve transportation conditions was an important consideration in the decision-making process. One of the most obvious examples is the Three Gorges Dam in China, which was advocated eagerly by provinces on the upper reaches of the Yangtze River, like Chongqing and Sichuan Province. After the dam was finished, the river's water level rose by 175 m, allowing 10,000-t ships to reach Chongqing, which has substantially improved the transportation condition of this major city surrounded by mountains (Jackson and Sleigh 2000).

1.3 Environmental Impacts of Large Dams

Most large dams constructed throughout the world have emphasized the economic and social benefits to society while largely ignoring any long-term environmental impacts (McCartney et al. 2001). More recently, however, the environmental impacts of large dams on river ecosystems have received considerable attention and the cost of dams to human society has become a major concern. Dams are designed to alter the natural flow regime of rivers and, as such, they have profound impacts on natural river processes both upstream and downstream from the dam. The impacts of dams on river ecosystems are numerous, complex, and varied, some obvious and others more subtle, but all mostly have negative consequences.

The environmental impacts of dams on river ecosystems can be viewed within a hierarchical framework of first-, second-, and third-order impacts with upstream and downstream effects considered separately (Table 1.1) (Bergkamp et al. 2000;

Table 1.1 A hierarchical framework of upstream and downstream impacts of dams on river ecosystems (From Bergkamp et al. 2000; McCartney et al. 2001)

Location in relation to dam	Category of impact	Impact
Upstream	First-order impacts	Modification of thermal regime
		Accumulation of sediment in reservoir
		Changes in water quality
		Evaporation and greenhouse gases
	Second-order impacts	Changes in channel configuration
		Increased growth of plankton and periphyton
		Increased growth of aquatic macrophytes
		Reduced biomass and diversity of riparian vegetation
	Third-order impacts	Changes in distribution and abundance of invertebrate, fish, bird, and mammal populations
Downstream	First-order impacts	Changes in the timing, magnitude, and variability of daily, seasonal, and annual flows
		Changes in water quality
		Reduced sediment flows
		Alteration of channel, floodplain, and coastal delta morphology
	Second-order impacts	Change in plankton and periphyton assemblages
		Increased growth of aquatic macrophytes
		Change in riparian vegetation
		Change in channel, floodplain, and coastal characteristics
	Third-order impacts	Changes in distribution and abundance of invertebrate, fish, bird, and mammal populations
		Increased salinity of estuaries

McCartney et al. 2001). First-order impacts are the immediate abiotic effects on the hydrology, water quality, and sediment load of the river as a direct consequence of the dam. These are the key driving variables that lead to second- and third-order impacts resulting in long-term changes of the river ecosystem. Second-order impacts are the abiotic and biotic changes in ecosystem structure and primary productivity caused by first-order impacts that take place over several years. Third-order impacts are the long-term biotic impacts on higher trophic levels that result from the integrated effects of first- and second-order impacts. In general, the complexity of the ecosystem processes and their functions that are altered increase from first- to third-order impacts.

1.3.1 Impacts on Flow Regime

The construction of a dam impounds water and converts the river channel from a lotic to lentic system in the upstream area. Flow regulation by the dam alters the intensity, timing, and frequency of downstream flow patterns reducing river discharge and flow variability by increasing low flows and dampening high flows.

The overall effect is a disruption in the longitudinal and lateral hydrologic connectivity of the river (McCartney et al. 2001; Hu et al. 2008; Lajoie et al. 2007; Bergkamp et al. 2000; Stanford et al. 1996). Downstream impacts are specific to each dam and depend on the storage capacity of the dam relative to the volume of river flow combined with how the dam is operated (Bergkamp et al. 2000; McCartney et al. 2001). In addition, the local climate can impact flow regime patterns in dam area as well (Zhao et al. 2012a).

1.3.2 Impacts on Thermal Regime

The large mass of water impounded behind the dam tends to increase water temperatures upstream and modifies the thermal regime of the river downstream by changing the natural seasonal and short-term temperature fluctuations downstream (McCartney et al. 2001). Deep reservoirs will thermally stratify in summer with warm, less-dense water in the upper layer (epilimnion) and the colder, denser water in the bottom layer (hypolimnion). Colder than normal water is released in the summer and warmer than normal water is released in the fall, altering seasonal temperature patterns of the river downstream (McCartney et al. 2009). Reservoirs at higher latitudes can easily become stratified due to variable input of solar energy, while those near the equator are least likely to become stratified (Bergkamp et al. 2000; McCartney et al. 2001). However, even if there is no thermal stratification, water released from reservoirs is often thermally different from the natural flow in the river and native aquatic organisms must adapt, relocate, or perish (Bergkamp et al. 2000; McCartney et al. 2009).

1.3.3 Impact of Evaporation and Greenhouse Gas Emission

Reservoirs increase the total surface area of freshwater which leads to water loss through evaporation. The amount of water evaporated from a reservoir is associated with both the surface area of the reservoir and the climatic conditions which control potential evaporation, i.e., evaporation is greatest from reservoirs with large surface areas located in hot arid climates (McCartney et al. 2001). The loss of additional water from evaporation can affect not only the water utilization efficiency by human beings, but also the downstream ecosystem health, e.g., water salinity associated with evaporation in arid climates can damage downstream river ecosystems.

Emission of greenhouse gases, carbon dioxide (CO_2) and methane (CH_4), from reservoirs as a result of decomposition of submerged vegetation has become a more recent concern (Fearnside 1997). Greenhouse gas emissions vary with reservoir depth, water residence times, temperature, the influx of organic matter from the catchment, age of the reservoir, rates of primary production, and dam operation (McCartney et al. 2001). Shallow tropical reservoirs are the largest emitters of greenhouse gases (McCartney et al. 2001), with average flux rates of 3,500 mg CO_2 and 300 mg CH_4 m^2/day reported for Panama, Brazil, and the French Guyana (Saint Louis et al. 2000). Relatively small reservoirs emit little or no greenhouse gases and are "carbon neutral" whereas large reservoirs can emit large quantities of CO_2 and CH_4 (Bergkamp et al. 2000).

1.3.4 Impacts on Sedimentation

The velocity and turbulence of a river slows down as it reaches a reservoir, and sediment particles being transported by the river drop out of suspension. All reservoirs accumulate sediments and lose water storage capacity over time, but the rate at which this happens varies widely. Sediment transport and deposition varies among rivers and depends on physiographic characteristics of the drainage basin (size, geology, soils, topography, and vegetation) as well as land use activities within the catchment area and the size of the reservoir relative to the volume of river flowing into the reservoir (Bergkamp et al. 2000; McCartney et al. 2001). Due to sediment deposition, about 0.5–1 % of the storage volume of the world's reservoirs are assumed to be lost annually (Mahmood 1987), leading to reduced water storage capacity and hydro-energy generation potentials of the dams. Heavy metals, which are carried by sediments, tend to accumulate in the upstream inundation area of a dam (DelValls et al. 1998). Downstream of a dam, a reduction in sediment load can lead to increased erosion of river banks and beds, loss of floodplains, and degradation of coastal deltas as sediment carried downstream by the river is no longer being replaced by material from upstream (McCartney et al. 2001). Reservoir flushing, the selective release of highly turbid waters, is sometimes used to reduce in-reservoir sedimentation (Atkinson 1996) and to increase sediment loads in the downstream river.

1.3.5 Impacts on Landscape and Morphology

Dam construction can greatly impact the landscape of the catchment area and morphology of the river channel with resultant impacts on the hydrologic regime of the river (Zhao et al. 2012b). Land use and land cover change are some of the most obvious impacts of dam construction on the landscape (Ouyang et al. 2010).

Once the dam is operational, reservoir sedimentation progressively changes the reservoir storage capacity and basin substrate (Buttling and Shaw 1973), which subsequently influences both the character of the water discharged below the dam and the suspended load passing the dam (McCartney et al. 2001). For example, the average annual loss of storage capacity due to sedimentation within the Saalachsee Reservoir, Bavaria decreased from 5.6 % during the first 15 years of operation to 2.5 % over a 47-year period and the output of suspended sediment nearly doubled (Bauer and Burz 1981).

The extent of channel morphology change below a dam depends on the interaction of three factors: the degree of flow regulation, the resistance of the channel bed and bank materials to erosion, and the quantity and nature of downstream sediment sources (McCartney et al. 2001). The reduction in sediment load tends to scour the channel below the dam, but the regulated flow regime (reduced peak flows) will counter this effect by reducing the flow velocity of the river. Both of these processes are modified by the erodability of the channel banks and downstream sources of sediments (Galay 1983; Williams and Wolman 1985). Channel erosion and degradation typically occurs within the first several kilometers below the dam. Further downstream, increased sedimentation (aggradation) may occur because suspended material in the river water is deposited due to slower moving water resulting in a widening of the channel (McCartney et al. 2001).

In addition, changes in sediment transport can result in changes in floodplains and even coastal delta and coastline morphology hundreds of kilometers below a dam site. Dams can either increase floodplain deposition or decrease it by erosion depending on specific conditions (McCartney et al. 2001). In some cases, the reduced frequency of flood flows and the stable low flows associated with the dam operation may encourage vegetation encroachment that can stabilize new deposits, further trap sediments, and reduce floodplain erosion. In other cases, the increased peak flood flow associated with dam operation may increase channel bank erosion, resulting in the loss of floodplains (McCartney et al. 2001). In coastal areas, the reduction of sediment input invariably leads to increased degradation. For example, dam construction on the Nile River has resulted in erosion rates of its delta of up to 5–8 m/year, and in some locations it reached as much as 240 m/year (McCully 1996).

1.3.6 Impacts on Water Quality

Dams change not only the hydrologic regime but also the chemical, biological, and physical characteristics of the water reducing the water quality and health of the river ecosystem (Ligon et al. 1995; Poff and Hart 2002; Hu et al. 2008). The quality of water stored behind a dam can be very different from the river water that flows into the reservoir. Impacts on water quality vary with the size of the dam, its location in the river system, its geographical location with respect to altitude and latitude, the detention time of the water, and the sources of the water (Bergkamp et al. 2000; McCartney et al. 2001). Large influxes of nutrients into reservoirs can occur as a consequence of both anthropogenic inputs within the water basin (e.g., fertilizer and sewage) and from the biological release of nutrients from flooded vegetation and soil (McCartney et al. 2001). As a result, eutrophication can occur resulting in water blooms of blue-green algae which can deplete oxygen in the deep water layer (Zakova et al. 1993).

Heavy metals originate from both natural and anthropogenic sources and can accumulate in the sediments of reservoirs (Shine et al. 1995; Kidd et al. 2007) impacting the aquatic food chain and leading to serious human health hazards

(Loska and Wiechuła 2003). A study of the Mawan Dam on the Lancang River (Upper-Mekong) measured high concentrations of As, Cd, Cr, Cu, Pb, and Zn in sediments that exceeded either the threshold effects level (TEL) of the US National Oceanic and Atmospheric Administration (NOAA) or Effects Range-Low (ERL) of the Canadian Sediment Quality guidelines (Zhao et al. 2012c).

Changes in water quality of the reservoir will also impact water quality downstream. Water released from a thermally stratified reservoir may change the natural temperature regime of the river, which impacts in-stream biota (Petts 1984), because it influences many important physical, chemical, and biological processes (McCartney et al. 2001). Different water release management practices of a stratified reservoirs will differentially change the water quality as water released from near the surface is generally well-oxygenated, warm, nutrient-depleted water whereas water released from near the bottom is cold, oxygen-depleted, nutrient-rich water (Zakova et al. 1993).

In arid climates, the salinization of water due to increased evaporation can be particularly problematic for downstream river ecosystems and floodplain wetlands (McCartney et al. 2001). Elevated salinity also will affect aquatic organisms (Hart et al. 1991).

1.3.7 Impacts on Aquatic Organism

The changes of hydrology, sedimentation, thermal regimes, and water quality associated with dam construction and operation can impact the composition and biomass of aquatic organisms including phytoplankton, periphyton, macrophytes, and fish in the food web. Upstream of a dam, the impoundment creates an ideal habitat for phytoplankton, but a less suitable habitat for periphyton and rooted macrophytes depending on depth, temperature, light penetration, and the nature of the substrate (McCartney et al. 2001). Changes in the composition of aquatic organisms can significantly affect others, e.g., blooms of phytoplankton and floating plants (e.g., water hyacinth) reduce light penetration and deplete oxygen when they decompose, and thus adversely affect other species (Joffe and Cooke 1997).

Downstream of a dam, primary production of phytoplankton, periphyton, and macrophytes are affected by changes in flow characteristics, water chemistry, turbidity, and thermal regimes. The flood mitigating characteristic of dams tends to promote the maintenance of higher than natural plankton populations within regulated rivers, by both sustaining populations released from the reservoir and promoting conditions for plankton development. Algal growth occurs in the channel immediately downstream from dams because of the nutrient loading from the reservoir releases (McCartney et al. 2001). Algal biomass was found to be up to 30 times greater downstream as compared to an upstream reference site with a very different species composition (Valentin et al. 1995). Flow regulation by a dam will lead to an increase in bed stability downstream, creating new habitat for the growth of aquatic plants. For example, flow regulation has allowed the rapid development of rooted plants (*Panicum repens* and *Phragmites mauritanus*) within the Zambezi River since the creation of Lake Kariba (Jackson and Davies 1976).

Fish are the most sensitive organisms affected by dam construction and flow regulation (Marchetti and Moyle 2001). Changes in the flow regime, water quality, river continuum, channel morphology, habitat conditions, and food webs may benefit some species, but they generally have an adverse effect on the majority of native species. Migratory fish are particularly susceptible due to the disruption of longitudinal connectivity by the dam (Wu et al. 2003). Dams can fragment and alter habitats (Wu et al. 2003; Wozney et al. 2011) to the point where native fish species are extirpated, allowing the invasion of exotic fish species and increasing biotic homogenization (Mckinney and Lockwood 1999; Olden et al. 2006; Olden and Rooney 2006). It is estimated that half the fish stocks endemic to the Pacific Coast of the United States have been lost in the past century, to a large extent because of dam construction (Chatterjee 1997). Upstream of the dam, the changes in flow patterns, and water temperature potentially change the feeding and spawning habitats of fish (Tiffan et al. 2002). Downstream of dams and fish populations change remarkably due to blockage of migration routes; disconnection of the river and floodplain; and changes in flow regime, physiochemical condition, primary production, and channel morphology (McCartney et al. 2001).

1.3.8 Impacts on Riparian Vegetation

Large dams can significantly change shoreline and riparian vegetation in both the impoundment region and downstream reaches (New and Xie 2008), even though the critical environment processes and diverse habitats for flora and fauna are supported by the riparian ecosystems (Beauchamp et al. 2007; Mallik and Richardson 2009). A number of studies have demonstrated that riparian vegetation can be remarkably impacted by damming, leading to habitat heterogeneity, declines in species richness and native species, and exotic species invasion (Nilsson and Svedmark 2002; Stave et al. 2005; Tealdi et al. 2011). Short-term vegetation changes are different from the long-term response due to riparian succession after dam begins operation (Nilsson et al. 1997).

Upstream of the dam, the largest impact on riparian vegetation is biomass submergence, leading to loss of plant biodiversity and primary production. Moreover, variation in the water levels of reservoirs can negatively affect vegetation in the immediate vicinity of the reservoir. For example, in Sweden, the pattern of water-level fluctuations regulated by a dam was not synchronized with the natural regime so that the riparian vegetation cover was extremely sparse (Nilsson and Jansson 1995).

Flow regulation below a dam has an adverse affect on plant species adapted to pulse-stimulated habitats. High discharges can retard the encroachment of true downstream terrestrial plants, while dams can disrupt plant reproduction and allow the encroachment of upland plants previously prevented by frequent flooding (McCartney et al. 2001). For example, *Acacia xanthophloea* is disappearing from the Pongolo system below Pongolapoort Dam, South Africa as a result of miss-timed floods (Furness 1978). However, given sufficient time after dam construction forest types more characteristic of unflooded upland areas may replace the original riparian forest vegetation.

1.3.9 Impacts on Biodiversity

Dams are believed to be not only a major cause for a decline in freshwater biodiversity, particularly the expiration of native fish species, the invasion of exotic fish species, and the decrease in fish beta-diversity, but also a major driver for the loss of terrestrial biodiversity along riparian zones. Dam construction can alter the structure and pattern of both terrestrial and aquatic habitats (Wu et al. 2004; Choi et al. 2005; Stromberg et al. 2007) resulting in the loss of biodiversity in river basins worldwide (Table 1.2). Riparian corridors are particularly important for many birds and terrestrial animals (Naiman et al. 1993), and the creation of reservoirs and inundation of ecosystems upstream will inevitably lead to the loss of these habitats (McCartney et al. 2001). Furthermore, the disruption of seasonal water flow regimes for flood control impacts downstream habitats putting the mammals and birds that use them at risk (McCartney et al. 2001; Nilsson and Dynesius 1994).

1.4 Social Impacts of Large Dams

Large dams significantly impact humans worldwide. Social impacts brought by dam construction generally occur in agricultural activities, economic activities, and local culture. Chapter 4 will narrow the scope to specific study areas in China, and analyze the impacts on local communities using more detailed information and a stringent analytical framework.

1.4.1 Impacts on Agriculture

Large dam projects have multiple negative impacts on agriculture at different scales, but the most direct and significant is the inundation of farmland. The conditions of the land along riverbanks are generally ideal for labor-intensive agriculture: relatively flat, easy access to irrigation water, and fertile. In one of our case studies in Chap. 2, Yunnan China, most of the highly productive paddy fields are along riversides. But since these fields are close to the river and at approximately the same elevation, most of them will be submerged by reservoirs once the dams are completed (Wang et al. 2013a).

Table 1.2 Rivers at risk—summary of threats to the biodiversity in the major river basins worldwide

Basin name	Countries within basin	Large dams	Types of risk
Yangtze	China	46	Large basin under stress from population pressures. Loss of habitat threatens bird species as well as endangered Yangtze River dolphin
La Plata	Argentina, Bolivia, Brazil, Paraguay, and Uruguay	27	River basin with high biodiversity; threats to Pantanal and other internationally important wetlands
Tigris-Euphrates	Turkey, Iraq, Syria, Iran, and Jordan	26	Arid basin; potential for conflicts over water withdrawal between Turkey and downstream countries
Salween	China, Myanmar, Thailand	16	Relatively pristine river with high biodiversity values; serious concerns about human rights abuses in Myanmar
Kizilirmak	Turkey	15	Small heavily fragmented basin; Ramsar site located in Delta
Ganges	India, Nepal, China, Bangladesh	14	Endangered Ganges River dolphin; Sundarbans mangroves in delta
Tocantins	Brazil	12	Relatively developed river basin; further dam development and improved navigation will exacerbate degradation for use of farmland
Amazon	Brazil, Peru, Bolivia, Colombia, Ecuador, Venezuela, Guyana, Suriname, Paraguay, and French Guyana	11	One of the most important basins for biodiversity; lower dams may affect coastal areas
Mekong	Thailand, Laos, China, Cambodia, Vietnam, and Myanmar	11	Basin with high biodiversity and very productive fisheries; droughts, and low water levels are current threats
Brahmaputra	China, India, Bhutan, Bangladesh	11	High biodiversity in upstream areas; high population pressure in delta
Zhu Jiang (Pearl River)	China, Vietnam	10	Highly developed basin; some important sites for amphibians
Danube	Germany, Austria, Slovakia, Hungary, Croatia, Serbia & Montenegro, Romania, Bulgaria, Moldova, Ukraine	8	68 Ramsar listed sites as well as UNESCO biosphere reserve in delta
Huang He (Yellow River)	China	8	River basin with severe water shortages; 4 endemic bird areas, 1 Ramsar site

(continued)

Table 1.2 (continued)

Basin name	Countries within basin	Large dams	Types of risk
Kura-Araks	Azerbaijan, Iran, Georgia, Armenia, Turkey	8	Biodiversity hotspot with 4 Ramsar sites, 1 Endemic Bird Area and 21 IBAs
Yesil-Kelkit	Turkey	8	Breeding populations of many birds in Delta
Büyük Menderes	Turkey	7	River delta protected as National Park, protected bird areas
Çoruh	Turkey	7	Fast flowing river with significant tourist industry based on rafting
Susurluk	Turkey	7	N/A
Ebro	Spain and Andorra	6	High economic importance of delta; important bird sites
Indus	Afghanistan, Pakistan, India, and China	6	Endangered Indus River Dolphin

Source: Adapted from World Wildlife Foundation (WWF 2005)

As a consequence of farmland inundation, new farmland needs to be cleared to compensate for the losses incurred by local farmers. These newly cleared lands, however, are in most cases inferior, since previously they were not arable. Problems relevant to newly cleared farmland include higher or lower temperatures, which may be inappropriate for the growth of certain crops; lower productivity, because it takes raw land several years before it can produce high yields; difficulties in irrigation, since the distance from the river is often much greater than before, and major investments in irrigation equipment are required to obtain the same amount of water as before; and some newly cleared lands have steeper slopes than before, which increases soil erosion. In addition, their relatively poorer conditions also force farmers to use more synthetic fertilizers, which can increase pollution in water bodies and ecosystems (Wang et al. 2013a; Wang 2012).

1.4.2 Impacts on Local Livelihood

Large dam projects submerge large areas of farmland, cause thousands of people to be relocated, and therefore have substantial impacts on local livelihoods. According to the World Commission on Dams (WCD 2000), 40–80 million people were relocated worldwide between 1950 and 2000 because of large dam construction.

The direct impacts of large dams on local people include the loss in houses, farmlands, forests, and other resources that determine a farmer's material wealth. To offset the losses, relocated people are sometimes given new houses and newly cleared farmland, while in other cases, their material wealth losses are addressed through monetary compensation programs. If only material wealth loss is considered, relocated people in many cases are fairly compensated, or in some cases even overcompensated (Wang et al. 2013a).

However, dam projects and associated relocation programs also have various indirect impacts on local people. First, many relocation schemes fail to consider the possible negative effects that might arise from moving to a new and different environment. People rely on specific skills to make a living, but the effectiveness of these skills can be impaired when moved to a totally new environment. Second, the social network supporting the community is probably disrupted in the process of relocation. For example, in a totally new environment, people might need to develop a new network for buying materials and selling products. Third, the inundation of a person's hometown can lead to significant psychological problems, which have occurred in some communities along the Upper-Mekong River (Zhao et al. 2012c; Wang et al. 2013a).

1.4.3 Impacts on Social Equity

The relocation of people due to dam construction not only impacts a farmer's wealth, but it also initiates a process of wealth reallocation. This may broaden the gap in relative wealth at different scales during relocation, which may lead to social inequity problems.

At the household level, relocation often drives people to find new and diverse employment opportunities, instead of continuing agricultural activities similar to those they did in their original village. Some villagers may be relocated to urban areas to work in factories or companies, while others may open their own small businesses. The result of this change in occupations is that some people end up being better off and some others become worse off. This new diversity widens the gap in wealth status among households, which may lead to social problems in the community (Fu and He 2003; Wang 2012).

The wealth status of entire villages also changes differentially because of different compensation policies and the different opportunities that arise from a specific dam project. Hence, compensation policies can vary substantially over time and in different regions. For instance, compensation standards for dam-induced relocation in China quadrupled in 2006 in response to new regulations, which caused large discrepancies in compensation between communities moved before and after this date (Wang et al. 2013a; Wang et al. 2013b). The opportunities available to relocated people also vary greatly. For example, some villages are moved close to an urban area with a convenient transportation system, which can stimulate the local economy. In contrast, other villages are moved to relatively isolated areas with limited access to services and very poor living conditions (Wang 2012).

At an even larger scale, such as the county or province level, the distribution of costs and benefits associated with the dam project may also be uneven. Regions close to a dam project likely will suffer most severely from its environmental and social impacts, but may not benefit greatly from the dam. For example, China's

Three Gorges Dam controls flooding in downstream provinces thousands of kilometers away, and generates electricity, which is fed into national grid, but only a small portion of these benefits reach local people in the dam-affected areas (Jackson and Sleigh 2000). A more detailed analysis of social inequities related to dam construction will be covered in Chap. 4.

1.4.4 Impacts on Culture

The impacts of dam construction on local culture can occur in multiple ways. First, the reservoir formed by the dam could inundate historical heritage sites along the river, since many ancient civilizations originated in river basins and left abundant historical structures. Second, dams are generally built in mountainous areas, which in many cases are the homes of ethnic minority groups with very diverse cultures. The relocation of these groups may disrupt their traditional lifestyles, causing the loss of many unique cultural traditions and intangible features of their heritages.

The Aswan Dam in Egypt flooded the original sites of many historical monuments, including Abu Simbel, Philae, Kalabsha, and Amada. The UNESCO and the Egyptian government moved many of these structures to higher ground above the reservoir to mitigate the loss of these historical resources (Hassan 2007). China's Three Gorges Dam also affected a large area rich with historical heritage. The most famous one is the "Mo ya shi ke," a place where huge Chinese calligraphy was carved along a cliff dozens of meters above the Yangtze River. The Chinese government moved whole pieces of rock about 50 m up the cliff so that this spectacular scenery remained visible for long distances after the reservoir filled (Childs-Johnson and Sullivan 1996).

The relocation of ethnic minority groups sometimes forces them to change lifestyles, therefore losing their traditions. Some ethnic groups are merged into another ethnic group's culture and have to adapt to the new environment by changing language and/or abandoning some of their old ways of living. For example, along the Upper-Mekong River many ethnic minorities no longer speak their own language after moving into areas dominated by Han, and they often lack appropriate space to celebrate some of their traditional festivals, therefore gradually abandoning these cultural events (Wang et al. 2013a).

1.5 Debates Around Large Dams

Debates about large dams have intensified in recent decades, and they generally fall into three types of concerns: economic feasibility, environmental sustainability, and social equity. This section briefly summarizes the different points of views commonly expressed about these concerns based on a series of key literatures on large dam issues (WCD 2000; Sadler et al. 2000; Magee 2006; World Bank 2009; Mertha 2008; Scudder 2005; Pan and He 2000).

1.5.1 Economic Feasibility

There are many proponents of dam construction, who mainly hold the opinion that building dams will generate opportunities for economic and social development, both at local and national levels. Their arguments include:

1. Dam construction projects will generate employment opportunities for local people and stimulate local economies because the arrival of engineers and construction workers will generate demands for various services.
2. The infrastructure that must be constructed before the dam project begins, such as roads and bridges will enhance the ability of local communities to access other regions, and sell their products across a larger market area.
3. After the dams come into operation, the reservoirs could provide irrigation water for farmland, stimulate a beneficial fishery, and attract tourists to the region because of their "lake-like" sceneries and recreational opportunities.
4. The navigation conditions for upstream regions will be improved, therefore facilitating trade and commerce activities.
5. Hydropower dams generate cheaper and cleaner electricity than those produced by fossil fuel combustion. Cheap electricity could reduce the costs of many industries and benefit the economy in general, and the profit from electricity could increase local government revenues and stimulate local economy.

However, the opponents to large dams argue that many dams actually fail to achieve their original economic goals. Many may not even recover the cost of their construction, nor the secondary costs to the environment and local communities. Their major arguments are as follows:

1. Many large dam projects overrun their budgets significantly, due to underestimating the technical difficulties, the compensation costs associated with the relocation processes, and changes in external conditions, such as the costs of labor and/or construction materials. Therefore, some large dams that are considered economically feasible actually end up failing to recover their total costs.
2. All large dams, regardless of their original purposes, have a problem of shortfall in designed goals to a certain extent. According to WCD's report (2000), most of the irrigation dams and water supply dams fail to achieve their designed goals, while hydropower dams are more likely to fulfill their targets.
3. Most large dam projects do not consider the externalized social and environmental costs in determining their total costs; therefore, the cost/benefit calculations for these projects are problematic. For example, large dams could harm a downstream fishery, but this financial loss is seldom incorporated into the actual cost of the dams during the decision-making process, especially when the downstream area is in a different political region.
4. There exist many alternatives to large dams, which may have much smaller economic, social, and environmental costs. Water-saving agricultural techniques have a great potential to reduce agricultural water demand. For example, micro-irrigation methods, such as sprinkler and drip systems, have already proved their

ability to save water. Renewable energy sources, including biomass, wind, solar, geothermal, and ocean tidal, only constitute about 1.5 % of the electricity generation around the world. If the use of renewable energy were increased greatly, the need to construct of hydropower dams might decline substantially.

1.5.2 Environmental Sustainability

Some opponents to dam construction argue that building large dams is an environmentally unsustainable solution for generating electricity, flood control, and irrigation for the following reasons:

1. The lifespan of a large dam is limited. Because of problems like sedimentation, most dams lose part or all their functions after about a hundred years. However, the impacts on the environment are irreversible and perpetual. Therefore, the short-term benefits of the large dams do not offset their long-term negative impacts.
2. Large dams alter the conditions of river systems so dramatically that many species cannot adapt appropriately, and therefore are likely to go extinct. According to various studies, dam construction is among the most harmful human-caused reasons for biodiversity loss.
3. The impacts of dam construction on the environment could act at different spatial scales, and therefore are very difficult to assess. Many studies focus only on the local impacts, but dams may also affect the regional climate and geological conditions, such as triggering earthquakes, and could even contribute to global climate changes because of increased greenhouse gas emissions.

The proponents of dam construction do not necessarily deny these negative impacts on the environment, but often argue that they are exaggerated. They note that ecosystems have the ability to adapt to the changes, and that there are many technological approaches that can be applied to mitigate the negative impacts. In sum, they argue that there is still an overall advantage to society to building large dams, as their positive benefits outweigh the negative impacts.

1.5.3 Social Equity

Opponents of dams argue that dam construction could generate many social inequity problems, mostly during the cost and benefit distribution and the resettlement processes. Their main arguments are as follows:

1. The distributions of costs and benefits of dam construction are uneven. The electricity generated from hydropower dams is in most cases fed into national grids, but local communities bear most of the negative impacts from dam construction. Most of the benefits arising from dams do not reach the local people.

2. The construction of dams often, if not always, involves large-scale relocation of local people. Relocated people lose their houses, land, and many other resources during the relocation, but are generally not fairly compensated. Most studies show that the dam-induced relocated people are generally adversely impacted economically and socially.
3. Local minorities and indigenous groups are underrepresented in the decision-making process, and are often unfairly treated in dam projects. Large dams are usually built in mountainous areas, which in many cases are the homes of ethnic minority groups. Statistics show that the percentage of ethnic minorities among all people affected by the dam projects is much higher than the percentage of the minorities in the whole population. These minorities generally lag in social and economic development, and are therefore more vulnerable to any changes brought about by dams.

The proponents of dams argue that better compensation policies that take regional benefits and costs and the rights of local people into consideration would resolve these social inequity problems. They suggest the following principles be considered during the decision-making process:

1. It is critical to make sure that all stakeholders involved in and impacted by a dam construction project participate in the decision-making process, and the entire process should be transparent to all stakeholders and even to society as a whole. Central and local governments, local communities, companies, scholars, and non-government organizations (NGOs) should have opportunities to express their concerns and opinions before the start of a dam project.
2. Better design in compensation policies that take various losses of relocated people into consideration is needed to mitigate the impacts to the minimum level. When relocation of people and communities is unavoidable in dam projects, reasonable compensation policies are critical to ensure the standard of living and the development opportunities of the relocated people are not greatly sacrificed.
3. The rights of the indigenous people and ethnic minority groups should be given particular attention and protected. Their interests should be considered during the decision-making process even though they may lack the ability to formally defend their own rights.

Because of the complexity and wide-range impacts of large dam construction, the ideas held by diverse stakeholders are often confounding and opposite, and make this issue a particularly difficult dilemma to address by modern societies worldwide. The following chapters will focus on China, a country that leads the world in the construction of large dams while also facing significant sustainable challenges arising from rapid and unprecedented economic development. By analyzing environmental and social impacts of large dams in different watersheds in China, and introducing frameworks to assess these impacts, we hope to provide helpful insights into the decision-making process for dam construction in China and beyond.

References

Atkinson E (1996) The feasibility of flushing sediment from reservoirs
Bauer F, Burz J (1981) Dar Einfluss der Festoffuhrung alpiner Gewasser auf die Stauraumverlandung und Flussbetteintiefung. Wasserwirttschaftliche Mitteilungen 4:114–121
Bawa KS, Koh LP, Lee TM, Liu J, Ramakrishnan PS, Yu DW, Zhang Y-P, Raven PH (2010) China, India, and the environment. Science 327:1457–1459
Beauchamp VB, Stromberg JC, Stutz JC (2007) Flow regulation has minimal influence on mycorrhizal fungi of a semi-arid floodplain ecosystem despite changes in hydrology, soils, and vegetation. J Arid Environ 68:188–205
Bednarek AT (2001) Undamming rivers: a review of the ecological impacts of dam removal. Environ Manage 27:803–814
Bergkamp G, McCartney M, Dugan P, McNeely J, Acreman M (2000) Dams, ecosystem functions and environmental restoration. Thematic review II. 1 Prepared as an input to the World Commission on Dams, Cape Town
Buttling S, Shaw T (1973) Predicting the rate and pattern of storage loss in reservoirs. Proceedings of the 11th International Congress on Large Dams, Madrid, Spain. International Commission on Large Dams, Paris, vol 1, pp 565–580
Chatterjee P (1997) Dam busting. New Scientist 154:34–37
Childs-Johnson E, Sullivan LR (1996) The Three Gorges Dam and the fate of China's southern heritage. Orientations 27:55–61
Choi SU, Yoon B, Woo H (2005) Effects of dam-induced flow regime change on downstream river morphology and vegetation cover in the Hwang River, Korea. River Res Appl 21:315–325
Delvalls TÁ, Forja J, González-Mazo E, Gómez-Parra A, Blasco J (1998) Determining contamination sources in marine sediments using multivariate analysis. Trends Anal Chem 17:181–192
European Environmental Agency (2010) Reservoirs and dams. http://www.eea.europa.eu/themes/water/european-waters/reservoirs-and-dams
Fearnside PM (1997) Greenhouse-gas emissions from Amazonian hydroelectric reservoirs: the example of Brazil's Tucuruí Dam as compared to fossil fuel alternatives. Environ Cons 24:64–75
Fu B, He Y (2003) The effect on the emigration's income and reservoir area ecology in caused by farmland change in reservoir area of Manwan Hydropower Station. Territory & Natural Resources Study
Furness HD (1978) Ecological studies on the Pongola River floodplain. Working Document IV, Workshop on man and the Pongolo floodplain. C.I.S.R., Pietermaritzburg, No. 14/106/7C
Galay V (1983) Causes of river bed degradation. Water Resour Res 19:1057–1090
Government of Beijing (1981) Beijing Government Document [1981] No. 124. http://210.73.64.113/Govfile/front/content/11981124_0.html
Hart BT, Bailey P, Edwards R, Hortle K, James K, McMahon A, Meredith C, Swadling K (1991) A review of the salt sensitivity of the Australian freshwater biota. Hydrobiologia 210:105–144
Hassan F (2007) The Aswan High Dam and the International Rescue Nubia Campaign. Afr Archaeol Rev 24:73–94
Hu W-W, Wang G-X, Deng W, Li S-N (2008) The influence of dams on ecohydrological conditions in the Huaihe River basin, China. Ecol Eng 33:233–241
Jackson PBN, Davies BR (1976) Cabora River in its first year: some ecological aspects and comparisons. Rhodesian Sci News 10:128–133
Jackson S, Sleigh A (2000) Resettlement for China's Three Gorges Dam: socio-economic impact and institutional tensions. Comm Post-Commun Stud 33:223–241
Joffe S, Cooke S (1997) Management of water hyacinth and other invasive aquatic weeds. Issues for the World Bank. Washington, DC. World Bank internal report
Kidd P, Dominguez-Rodriguez M, Diez J, Monterroso C (2007) Bioavailability and plant accumulation of heavy metals and phosphorus in agricultural soils amended by long-term application of sewage sludge. Chemosphere 66:1458–1467
Lajoie F, Assani AA, Roy AG, Mesfioui M (2007) Impacts of dams on monthly flow characteristics. The influence of watershed size and seasons. J Hydrol 334:423

Ligon FK, Dietrich WE, Trush WJ (1995) Downstream ecological effects of dams. BioScience 45:183–192
Liu J, Diamond J (2005) China's environment in a globalizing world. Nature 435:1179–1186
Longman J (2008) Dams are rejected in America as too destructive. Yet they are still promoted in Latin America. Why? Newsweek. http://www.thedailybeast.com/newsweek/2008/09/12/generating-conflict.html
Loska K, Wiechuła D (2003) Application of principal component analysis for the estimation of source of heavy metal contamination in surface sediments from the Rybnik Reservoir. Chemosphere 51:723–733
St. Louis VL, Kelly CA, Duchemin É, Rudd JW, Rosenberg DM (2000) Reservoir surfaces as sources of greenhouse gases to the atmosphere: a global estimate. BioScience 50:766–775
Magee DL (2006) New energy geographics: powershed politics and hydropower decision making in Yunnan, China. PhD Dissertation, University of Washington
Mahmood K (1987) Reservoir sedimentation: impact, extent, and mitigation. Technical paper. International Bank for Reconstruction and Development, Washington, DC
Mallik AU, Richardson JS (2009) Riparian vegetation change in upstream and downstream reaches of three temperate rivers dammed for hydroelectric generation in British Columbia, Canada. Ecol Eng 35:810–819
Marchetti MP, Moyle PB (2001) Effects of flow regime on fish assemblages in a regulated California stream. Ecol Appl 11:530–539
McCartney M, Sullivan C, Acreman M (2001) Ecosystem impacts of large dams. Background Paper No. 2. Prepared for IUCN/UNEP/WCD
McCartney M, Gichuki F, Nguyen-Khoa S, Kodituwakku D (2009) Living with dams: managing the environmental impacts. Water Policy 11:121–139
McCully P (1996) Silenced rivers: the ecology and politics of large dams. Zed Books, London, p 350
McKinney ML, Lockwood JL (1999) Biotic homogenization: a few winner replacing many loser in the nest mass extinction. Tree 450–453
Mertha A (2008) China's water warriors: citizen action and policy change. Cornell University Press, Cornell
Naiman RJ, Decamps H, Pollock M (1993) The role of riparian corridors in maintaining regional biodiversity. Ecol Appl 209–212
New T, Xie Z (2008) Impacts of large dams on riparian vegetation: applying global experience to the case of China's Three Gorges Dam. Biodivers Conserv 17:3149–3163
Nickum JE (1998) Is China living on the water margin? China Quart 156:890–898
Nilsson C, Dynesius M (1994) Ecological effects of river regulation on mammals and birds: a review. Regul Rivers: Res Manage 9:45–53
Nilsson C, Jansson R (1995) Floristic differences between riparian corridors of regulated and free flowing boreal rivers. Regul Rivers: Res Manage 11:55–66
Nilsson C, Svedmark M (2002) Basic principles and ecological consequences of changing water regimes: riparian plant communities. Environ Manage 30:468–480
Nilsson C, Jansson R, Zinko U (1997) Long-term responses of river-margin vegetation to water-level regulation. Science 276:798–800
Olden JD, Rooney TP (2006) On defining and quantifying biotic homogenization. Glob Ecol Biogeogr 15:113–120
Olden JD, Poff NL, Bestgen KR (2006) Life-history strategies predict fish invasions and extirpations in the Colorado River Basin. Ecol Monogr 76:25–40
Ouyang W, Hao F, Zhao C, Lin C (2010) Vegetation response to 30 years hydropower cascade exploitation in upper stream of Yellow River. Commun Nonlinear Sci Numer Simulat 15:1928–1941
Pan J, He J (2000) Large dams in China: a fifty-year review, China. WaterPower Press, Beijing
Petts GE (1984) Impounded rivers: perspectives for ecological management. Wiley, Chichester, p 326
Poff NL, Hart DD (2002) How dams vary and why it matters for the emerging science of dam removal. Bioscience 52:659–668
Sadler B, Verocai I, Vanclay F (2000) Environmental and social impact assessment for large scale dams, WCD Thematic Review V.2 prepared as an input to the World Commission on Dams. WCD, Cape Town

Scudder T (2005) The future of large dams: dealing with social, environmental, institutional and political costs. Earthscan, London

Shah Z, Kumar M (2008) In the midst of the large dam controversy: objectives, criteria for assessing large water storages in the developing world. Water Res Manage 22:1799–1824

Shine JP, Ika RV, Ford TE (1995) Multivariate statistical examination of spatial and temporal patterns of heavy metal contamination in New Bedford Harbor marine sediments. Environ Sci Technol 29:1781–1788

Stanford JA, Ward J, Liss WJ, Frissell CA, Williams RN, Lichatowich JA, Coutant CC (1996) A General Protocol for Restoration of Regulated Rivers. Regul Rivers: Res Manage 12:391–413

Stave J, Oba G, Stenseth NC, Nordal I (2005) Environmental gradients in the Turkwel riverine forest, Kenya: hypotheses on dam-induced vegetation change. For Ecol Manage 212:184–198

Stromberg JC, Lite SJ, Marler R, Paradzick C, Shafroth PB, Shorrock D, White JM, White MS (2007) Altered stream flow regimes and invasive plant species: the Tamarix case. Glob Ecol Biogeogr 16:381–393

Tealdi S, Camporeale C, Ridolfi L (2011) Modeling the impact of river damming on riparian vegetation. J Hydrol 396:302–312

The City of New York (2013) Watershed protection. http://www.nyc.gov/html/dep/html/watershed_protection/index.shtml. Accessed July 1, 2013

Tiffan KF, Garland RD, Rondorf DW (2002) Quantifying flow-dependent changes in subyearling fall Chinook salmon rearing habitat using two-dimensional spatially explicit modeling. N Am J Fish Manage 22:713–726

Valentin S, Wasson J, Philippe M (1995) Effects of hydropower peaking on epilithon and invertebrate community trophic structure. Regul Rivers: Res Manage 10:105–119

Vorosmarty CJ, McIntyre PB, Gessner MO, Dudgeon D, Prusevich A, Green P, Glidden S, Bunn SE, Sullivan CA, Liermann CR, Davies PM (2010) Global threats to human water security and river biodiversity (vol 467, pg 555, 2010). Nature 468:334

Wang P (2012) Social impact analysis of large dams: a case study of cascading dams on the upper-mekong river, China. Master's Thesis

Wang P, Lassoie JP, Dong S, Morreale SJ (2013a) A framework for social impact analysis of large dams: a case study of cascading dams on the Upper-Mekong River, China. J Environ Manage 117:131–140

Wang P, Wolf SA, Lassoie JP, Dong S (2013b) Compensation policy for displacement caused by dam construction in China: an institutional analysis. Geoforum 48:1–9

WCD (2000) Dams and development: a new framework for decision-making. World Commission on Dams

Williams G, Wolman M (1985) Downstream effects of dams on alluvial rivers. US Geol. Surv., Prof. Pap. (United States), 1286

World Bank (2009) Directions in hydropower

Wozney KM, Haxton TJ, Kjartanson S, Wilson CC (2011) Genetic assessment of lake sturgeon (Acipenser fulvescens) population structure in the Ottawa River. Environ Biol Fishes 90:183–195

Wu J, Huang J, Han X, Xie Z, Gao X (2003) Three-Gorges Dam—experiment in habitat fragmentation? Science 300:1239–1240

Wu J, Huang J, Han X, Gao X, He F, Jiang M, Jiang Z, Primack RB, Shen Z (2004) The three Gorges dam: an ecological perspective. Front Ecol Environ 2:241–248

WWF (2005) Rivers at risk: dams and the future of freshwater ecosystems. http://www.panda.org/downloads/freshwater/riversatriskfullreport.pdf

Xu J (1998) Naturally and anthropogenically accelerated sedimentation in the lower Yellow River, China, over the past 13,000 years. Geogr Ann 80:67–78

Yang T, Zhang Q, Chen YD, Tao X, Xu C-Y, Chen X (2008) A spatial assessment of hydrologic alteration caused by dam construction in the middle and lower Yellow River, China. Hydrol Process 22:3829–3843

Zakova Z, Berankova D, Kockova E, Kriz P, Mlejnkova H (1993) Investigation of the development of biological and chemical conditions in the Vir Reservoir 30 years after impoundment. Water Sci Technol 28:65–74

Zhang M, Jin Y (2008) Building damage in Dujiangyan during Wenchuan earthquake. Earthq Eng Eng Vibr 7:263–269

Zhao Q, Liu S, Deng L, Dong S, Yang J, Wang C (2012a) The effects of dam construction and precipitation variability on hydrologic alteration in the Lancang River Basin of southwest China. Stoch Environ Res Risk Assess 26:993–1011

Zhao Q, Liu S, Deng L, Dong S, Yang Z, Yang J (2012b) Landscape change and hydrologic alteration associated with dam construction. Int J Appl Earth Obs Geoinf 16:17–26

Zhao Q, Liu S, Deng L, Yang Z, Dong S, Wang C, Zhang Z (2012c) Spatio-temporal variation of heavy metals in fresh water after dam construction: a case study of the Manwan Reservoir, Lancang River. Environ Monit Assess 184:4253–4266

Chapter 2
Large Dams in China: An Overview of History, Distribution, and Case Studies

2.1 The History of China's Dam Construction

China has a long history of building dams and levees for flood control and irrigation. The earliest dam remains found so far indicate that it was built about 2,600 years ago, and some of the dams built more than 2,000 years ago are still functioning. These hydrologic projects contributed greatly to building a prosperous agricultural society in ancient China. But most of these ancient dams were relatively low, not meeting today's height definition for a large dam (>15 m).

According to the World Commission on Dams (2000), before 1950, there were 5,196 large dams worldwide, but only 22 were in China. However, the speed of dam construction in China increased so rapidly during the second half of the twentieth century that after 1982, China was building more dams than all other countries combined. The construction rate slowed during the 1990s, when there were 1,100–1,700 large dams under construction every year worldwide. Likewise, there were only 250–320 large dams under construction each year in China, but this still represented 20–25 % of the world's annual total. Another significant trend began in 1995, when India's building rate surpassed China's for the first time. The history of modern dam construction in China (i.e., post-1950) can be further examined by considering four separate time periods that reflect differing building rates (Pan and He 2000).

The first period was from 1950 to 1957, when large dam construction was just beginning in China. Strategic plans during this period focused on managing several hazardous rivers, including the Huai, Hai, and Huang. The heights of most dams constructed during this period were from 50 to 100 m, and their main purpose was to control major floods that were occurring in these river basins. Some of the famous dams built during this period include Guanting Dam (46 m) in Beijing, Foziling Dam (74.4 m) in Anhui Province, and most importantly, Sanmenxia Dam (106 m) in Henan Province, which will be presented as case study later in Chap. 4.

The second time period, 1958–1966, featured the nationwide, large-scale construction of infrastructure, including large dams. Different levels of political

P. Wang et al., *The Large Dam Dilemma: An Exploration of the Impacts of Hydro Projects on People and the Environment in China*, DOI 10.1007/978-94-007-7630-2_2,

districts actively participated in this effort, and the number of large dams in China increased dramatically. Some very large dams were constructed in this period, such as Liujiaxia Dam (147 m) in Gansu Province.

The third period was from 1967 to 1986. Because of the influence of political events, such as the Cultural Revolution, the rate of dam building was noticeably reduced. However, some of the important dams were constructed during this period, such as Gezhouba Dam on the Yangtze River in Hubei Province, which played an important role in the later construction of Three Gorges Dam. During this period, the quality and technology associated with building dams also advanced substantially.

The fourth period extended from 1987 to the present. China's successful economic development during this period not only provided financial support for large-scale infrastructure projects, but it also generated a huge demand for electricity, which has been a major stimulus for increasing hydropower projects nationwide. After long debates, the building of two of the world's largest dams, Xiaolangdi Dam in Henan Province and the Three Gorges Dam in Hubei Province, was initiated during this period. These will be discussed in greater detail in the case study section of this chapter.

The administration of reservoirs in China depends on their water-holding capacities (Pan and He 2000): reservoirs with capacities over 100 million m^3 are administrated by provincial or higher level agencies; those with capacities between 10 and 100 million m^3 are administrated by prefecture level agencies; those with 1–10 million m^3 are administrated at the county level; those from 100 thousand to 1 million m^3 are administrated at the town level; and reservoirs under 100 thousand m^3 are administrated at the village level. Hydroelectric plants have different administrative systems and will not be discussed here.

2.2 The Distribution of Large Dams in China

In the 1980s, the Chinese Government designed a long-term development plan for hydraulic projects, most of them focused on hydropower. The Ministry of Water Resources identified 13 hydropower bases in different river basins across China (Fig. 2.1) (China Hydropower Engineering Web 2011). Hydropower was considered a cleaner and cheaper energy source than fossil fuels, and this ministry proposed increasing the proportion of hydropower in the electricity structure of China, as coal-burning power plants currently contribute about 80 % of the nation's total electricity. This long-term hydraulic project plan covered almost all the large river basins in China (Fig. 2.1), and its implementation has been intensified due to the growing demand for energy as a result of rapid urbanization and economic development. Later sections of this chapter will provide case studies of some important dams, including the Three Gorges Dam, Xiaolangdi Key Water Control Project, and dams in Yunnan Province. More case studies are given in Chaps. 3 and 4, which focus on the environmental and social impacts of these dams, respectively.

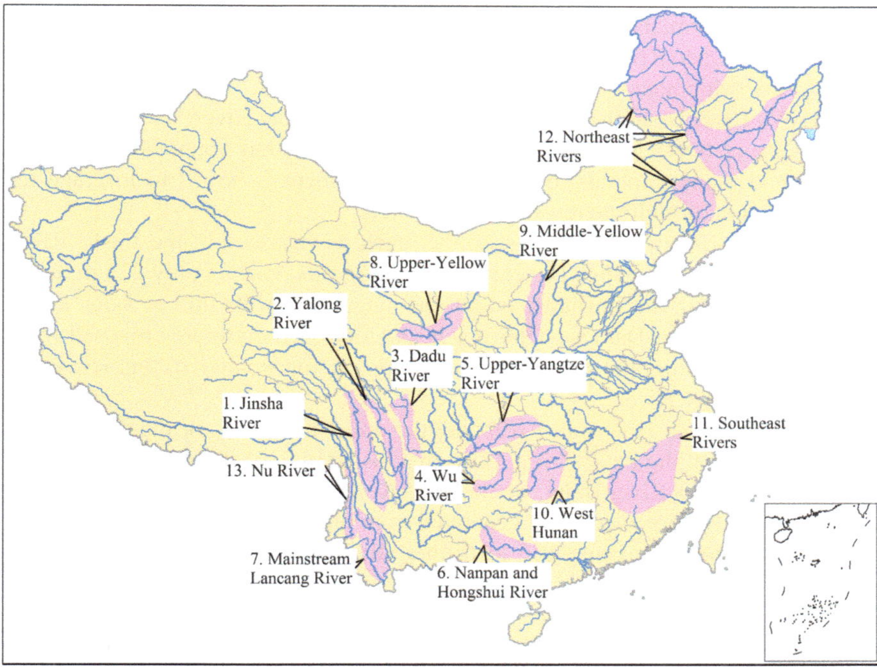

Fig. 2.1 Thirteen hydropower bases in China (Synthesized from data on China Hydropower Engineering Web (2011))

Each of these 13 hydropower bases has important characteristics that are summarized below (China Hydropower Engineering Web 2011). The numbers correspond to those provided in Fig. 2.1.

1. The Jinsha River is another name for the upper reach of the Yangtze River and is also one of the three parallel rivers in Yunnan Province. It has a large flow magnitude and drop in elevation, and therefore is potentially the largest source of hydropower in China. The feasible, installed capacity of this river alone is 63,830 MW, and there are dams either proposed or under construction on its middle and lower reaches; the upper reach of the Jinsha River has not yet been exploited.
2. The Yalong River is located in western Sichuan Province and is the largest tributary of the Jinsha River. Its middle and lower reaches run through deep valleys and have a significant elevation drop. The first hydropower station on the Yalong River, the Ertan hydropower station, had an installed capacity of 3,300 MW and was the largest hydropower station built during the twentieth century in China.
3. The Dadu River is a secondary tributary of the Yangtze River, but it has tremendous hydropower potential. Its elevation drop is more than 1,800 m, with a feasible, installed capacity of 23,480 MW. Because the estimated extent of

submerged farmland and villages is much less than that for other large dams in China, the Dadu River is considered to be very economically feasible for hydropower development.

4. The Wu River is another large tributary of the Yangtze River, with an installed capacity of 6,875 MW. Since land around the river has abundant deposits of coal and many minerals, hydropower development is hoped to leverage the exploitation of multiple resources in the area.
5. The Upper-Yangtze River refers to the reach downstream of the Jinsha River, from Yibing to Yichang. The elevation drop in this reach is only 220 m, much less than that in the Jinsha River, but the magnitude of its flow is much larger. The Three Gorges Dam is on this reach, and four other dams are proposed along this reach. Additional dams also are proposed for smaller tributaries of the Upper-Yangtze River.
6. The Nanpan and Hongshui Rivers are the upper tributaries of the third largest river in China, the Pearl River. They are close to the most developed region in China, the Pearl River Delta, where Hong Kong, Guangzhou, and Shenzhen are located. Therefore, electricity generated from these two rivers could be fed into the Guangdong Province grid to further stimulate economic growth in these major urban areas.
7. The Lancang River is another of the three parallel rivers in Yunnan Province, which is also the name of the upper reach of the Mekong River. Six dams were proposed for the upper reach of Lancang River, and another eight for the lower reach. The total installed capacity for the mainstream of the Lancang River is 21,370 MW. Dams on this river are discussed in detail in the case study parts of this chapter and Chap. 3.
8. There are 16 dams designed on of the Upper-Yellow River, with total, installed capacity of 14,155 MW. These dams are located in gorges in northwest China that have low population densities and less productive farmlands. Therefore, the economic impacts of these dams on local people are comparatively low.
9. The Middle-Yellow River runs through the longest stretch of gorges on this river. Since this reach is the source of the largest amount of sand along the river, dams built here would not only generate electricity, but also would control siltation in downstream dams, such as Sanmenxia and Xiaolangdi. Its estimated installed capacity is 6,092 MW.
10. There is a series of dams planned for rivers in western Hunan Province, which have great hydropower potentials, including the Yuan, Zi, and Li Rivers. It is proposed to build cascade dams on each of these three major rivers, with total installed capacity of 6,613 MW. Beside hydropower, they also will have the ability to control flooding in cooperation with dams on the mainstream of the Yangtze River.
11. The mountainous areas in the southeast provinces of Fujian, Zhejiang, and Jiangxi contain tremendous hydropower potential, with an estimated total installed capacity of 16,800 MW.

12. The northeast China hydropower base includes five major rivers, namely the Heilong, Mudan, Songhua, Yalu, and Nen. The total installed capacity is estimated to be 11,316 MW.
13. The Nu River is the third of the three parallel rivers in Yunnan Province. Its proposed installed capacity is 21,320 MW. Because of the protests arising from various NGOs as well as the Ministry of Environment Protection, no dams have been constructed to date on this river, and it is called the "last pristine major river in China." This will be discussed in the case study part of Chap. 4.

2.3 Case Studies of Large Dams in China

This section provides a series of case studies on representative large dams in China, including the Three Gorges Dam, Xiaolangdi Key Water Control Project, and Cascade Dams on the Lancang (Upper-Mekong) and Jinsha Rivers. The first two dams are key hydraulic projects built on the two largest rivers in China, the Yangtze and the Yellow, respectively. The other two cases are in Southwestern China, where there is the highest hydropower potential in China. The case studies in this section intend to provide a general sense of the physical, social, and environmental features of large dams in China, from the decision-making process to the various impacts of large dams. Chapters 3 and 4 will use more case studies and focus on the environmental and social impacts of large dams, respectively.

2.3.1 The Three Gorges Dam

2.3.1.1 Seventy Years of Decision-Making

The Three Gorges Dam was among the most contentious dams in China's hydraulic project development history. The original idea of building a large dam on the Yangtze River before it runs into the plains of Central China was proposed in 1919 by Sun Yat-sen, and in 1932 the government of Republic of China conducted the first investigation to identify potential sites for such a dam. However, at that time it was not economically viable or technologically feasible for the Chinese Government to build such a dam. In 1944, John Lucian Sovage, a dam expert from the US Bureau of Reclamation, investigated the Three Gorges area, and wrote the first scientific proposal for constructing the dam. In this proposal, the electricity generated by the dam surpassed the total demands of all seven provinces around the Three Gorges. China and the United States began to consider a collaborative effort to build the dam, but this plan was terminated in 1947 due to the Chinese Civil War (Xinhua News Agency 2003b).

After the founding of the People's Republic of China in 1949, the dam proposal was again brought into serious discussion, when MAO Zedong showed special interest in constructing the dam to harness the Yangtze River. Mao and many hydraulic engineers investigated the Upper-Yangtze many times and discussed optimal locations for the dam and its expected cost, lifespan, and possible difficulties that might be encountered during its construction and operation. However, the budget proposed by hydraulic engineers to Mao was still too large for the fragile national economy. Even so, Mao's desire to build this dam played an important political role in later decision-making processes (Childs-Johnson and Sullivan 1996; Xinhua News Agency 2003b).

In 1984 the Yangtze River Basin Committee proposed a plan to the Central Government, with a dam height of 150 m. However, the City of Chongqing was dissatisfied with this plan because the backwater of the reservoir would not reach Chongqing. Hence, the final dam height was changed to 185 m, with the water level rise of 175 m being equal to the elevation of the Chaotianmen Port in Chongqing, which would greatly improve the transportation conditions and commerce for this megacity. In 1992, after many years of debates and discussions, and being more confident in its economic abilities, the Chinese Central Government finally made the decision to build the dam.

Construction commenced in 1994, with the whole project divided into three stages. The first stage lasted from 1994 to 1997 and included preparing projects to divert the Yangtze River's flow and designing a temporary ship lock along the eastern bank. The second stage, 1998–2003, involved building the dam and installing generators on the east side and completing a permanent ship lock. The third stage, 2003–2009, involved finishing the dam and installing generators on the west side. After the completion of the project, the reservoir behind Three Gorges Dam currently is 600 km long, covering a total area of 10, 000 km^2 (Stone 2008; Xinhua News Agency 2003b).

2.3.1.2 Multiple Functions

Three Gorges Dam is a megaproject with multiple purposes. The first purpose is to generate electricity. Due to the huge reservoir capacity, it now has the ability to generate the greatest amount of electricity among all dams worldwide. The installed capacity is 18,200 MW, and the electricity from its power station is transported to 10 provinces, including Shanghai thousands of kilometers away (Xinhua News Agency 2003a; Jackson and Sleigh 2001; Stone 2008).

The second purpose is flood control. The downstream area of the Yangtze River has been vulnerable to flooding for thousands of years, which has been exacerbated more recently by deforestation in its upstream areas. The Yangtze's downstream plain is the most populated and developed region in China and three large floods during the twentieth century, 1931, 1954, and 1998 have caused thousands of deaths and staggering losses in capital. In 1992, right before the final decision was made to construct the dam, some experts stated the important function of flood control as follows: "If we don't consider the flood control of the Three Gorges Dam, then its costs will exceed the benefits; however, if we take the benefit of flood control into account, then the benefits of the dam will exceed the costs" (Xinhua News Agency 2003a). This statement illustrates the significance of the dam's flood control function. It was reported that the Three Gorges Dam reservoir has a flood storage capacity of 22.15 billion m^3 and could alone protect downstream areas from 100-year return-period floods. Furthermore, in collaboration with other flood control projects, the dam also would significantly reduce damages arising from 1,000-year floods (Xinhua News Agency 2003a; Wu et al. 2004).

The third purpose is for navigation. Flow in the upstream reaches of the Yangtze used to be rapid and narrow, but after dam construction, it became placid and wide. The water is also much deeper than before, which allows larger ships to reach upstream cities. For example, the Port of Chongqing can now accommodate large cargo ships in the 10,000-t class connecting commerce to downstream locations. This improvement in water transportation has significantly stimulated economic development in this megacity (Xinhua News Agency 2003a).

2.3.1.3 Social and Environmental Consequences

The Three Gorges Dam has caused various social and environmental impacts since its construction. It was estimated in official documents that after the completion of the relocation program in 2010, a total of 1,397,000 people had been relocated. The dam directly affected 20 counties in Chongqing City and Hubei Province; but many more counties and provinces were involved in the relocation project. As in the original plan, 25,000 people in affected counties in Hubei were relocated to other unaffected counties in the same province; 20,000 people in affected counties in Chongqing were relocated to other unaffected counties also in Chongqing, but 70,000 people were relocated to other provinces, specifically, Sichuan, 9,000; Jiangsu, Zhejiang, Shandong, Hubei, and Guangdong, 7,000 each; Shanghai and Fujian, 5,500 each; and Anhui, Jiangxi, and Hunan, 5,000 each. The actual numbers were generally much higher than those cited in the original plan; for example, Anhui Province actually accepted 8,094 residents from Chongqing rather than 5,000 as stated in the plan (Xinhua News Agency 2003a; Jackson and Sleigh 2000; Li et al. 2001).

After settling in new places, relocated people began their long and sometimes difficult adaption to a new environment. Lack of farmland and other resources made the lives of some relocated people extremely difficult, and it was also hard for many to merge into the local society. For example, it became top news in Suqian, Anhui Province when a local woman married a relocated man because they were discriminated against and this kind of marriage was very uncommon (People's Daily Online 2010). The undetermined social costs of the almost 1.4 million relocated people make the total cost of the Three Gorges Dam difficult to estimate.

The environmental impacts of the Three Gorges Dam are also substantial. The dam altered the Yangtze River's flow regime significantly, which extensively influenced water quality (Muller et al. 2008), terrestrial and aquatic biodiversity (Wu et al. 2004), and fisheries (Chen 2002). Records showed that after the construction of the dam, the concentration of nitrate almost doubled downstream, and the concentration of many heavy metal ions, such as Pb, Cu, Cd, and Cr, also significantly increased. These changes in water quality could threaten the health of downstream ecosystems. Water level in the reservoir increased by 175 m and isolated more than 100 mountaintops and ridges, turning them into islands. This fragmentation of habitats could greatly reduce the number of species. The dam has even larger impacts on aquatic biodiversity, such as the blocking of migratory pathways and fragmenting of aquatic habitats, which are particularly detrimental to many species, some of which are endangered. The fishery on the Yangtze River also suffers from the dam. Annual commercial harvest data show that after 2003–2005, which was the period of impoundment, the commercial harvest of four carp species and the number of drift-sampled carp eggs and larvae decreased dramatically.

2.3.2 Xiaolangdi Key Water Control Project

2.3.2.1 A Failure Precedent: Sanmenxia Dam

The Yellow River is the most hazardous river in China, and it is also a river with the world's largest concentration of sand. Due to serious ecological degradation and soil erosion in its middle reach, 1.35 billion tons of sand are deposited in the river every year during the flood season. In the lower reach where the river course becomes flat and the flow speed decreases, about a quarter of this sand accumulates along the riverbed, lifting it by 10 cm each year. After thousands of years of sedimentation, the riverbed of the lower reach is now on average 5 m higher than the surrounding land, and a 1,400 km long levee has been built and maintained to hold the water. Hence, flood control of the Yellow River has long been a key concern of the Chinese Government (Ministry of Water Resources 2009).

The Chinese Government decided to build a large dam in the gorge areas of the Yellow River in the 1950s to resolve the sediment accumulation and flood problems. Experts from the Soviet Union were invited to help design the dam. Those experts were experienced in large hydraulic projects in the Soviet Union, but had never worked on a river with such a high sand concentration. They finally chose Sanmenxia Gorge as the dam site. However, this proposal was strongly opposed by some Chinese experts who were familiar with the characteristics of the Yellow River, particularly Huang Wanli, who was a professor at Tsinghua University and widely respected for his knowledge about the management of the Yellow River. Huang argued that building a dam at Sanmenxia would block the sand from moving downstream and cause serious floods on the upstream plain area. Unfortunately, it was difficult to oppose experts from the Soviet Union at that time, so the proposal was approved and Huang suffered political persecution for more than 20 years (Tianjin E-North Netnews 2003; Tan and Liu 2003).

Sanmenxia Dam was complete in 1960. Unfortunately, only 1-year later Huang's prediction became true when 1.5 billion tons of sand blocked the Yellow River, lifting the riverbed of the Wei River, the Yellow River's largest tributary, by 40 m.

A large area of highly productive farmland on the Wei River plain was inundated, and almost half a million local people were forced to move to Gansu Province hundreds of kilometers away. Even though Sanmenxia Dam was overhauled several times, it continued to cause serious problems for the next 40 years. The failure of this dam was the result of suppressing different opinions during the decision-making process and the lack of consideration of how to reduce sediment accumulation. Having learned from these mistakes, the Chinese Government during the 1990s began planning for another large dam, Xiaolangdi Dam about 130 km downstream from Sanmenxia Dam, which would fulfill the unachieved goals of this earlier dam (Tan and Liu 2003; Tianjin E-North Netnews 2003).

2.3.2.2 The Historical Missions of Xiaolangdi Dam

The Xiaolangdi Key Water Control Project is located 40 km north to Luoyang City, Henan Province, and 130 km downstream of Sanmenxia Dam. It is at the last gorge mouth of the middle Yellow River, and controls an area of 69,420,000 km^2 of river basin, which is about 92 % of the entire Yellow River basin. The multiple purposes of Xiaolangdi Dam include flood control, sediment reduction, water supply, irrigation, and hydropower (Ministry of Water Resources 2009).

After the failure of Sanmenxia Dam, Xiaolangdi Dam was given high expectations to solve the serious problems related to the Yellow River. During the rainy season, the downstream levee becomes very fragile, and because of the continuing accumulation of sand on the riverbed, the levee needs to be strengthened and heightened year after year. A new problem emerged during the 1970s when heavy water consumption in the Yellow River watershed began causing water shortages downstream during the dry seasons. Therefore, four goals were set for Xiaolangdi Dam: (1) during the rainy seasons the levee would not break; (2) during the dry seasons the water course would not dry up; (3) water contamination would be kept at a low level; and (4) sand would not continue to accumulate on the downstream water course, and thus the height of the waterbed would not increase (Henan Xinhua News Agency 1997; Ministry of Water Resources 2009).

In April 1991 the National People's Council approved the Xiaolangdi project. Dam construction started in September 1991 and was complete at the end of 2001. The top of the dam is 281 m above sea level, and the maximum water level is 275 m above sea level. The total capacity of the dam reservoir is 12.65 billion m^3, which includes 7.55 billion m^3 for storing sand and other 5.1 billion m^3 for regulating water flow and generating electricity. The World Bank provided a loan of one billion US dollars for this project, as well as technical and management support (Ministry of Water Resources 2009; China.org.cn 2008).

After the completion of the dam, Xiaolangdi began to play an important role in river management. First, the massive reservoir could store water during the rainy seasons, thus reducing the risk of floods in the downstream areas. Before the Xiaolangdi Dam was built, the existing levee and reservoirs on the Yellow River could only provide protection from 60-year return-period floods, but now downstream areas are safe from 1,000-year floods (Liu and Xia 2004; Ministry of Water Resources 2009).

Second, the reservoir stores water during the rainy season, and discharges it during the dry season to meet water consumption demands downstream, particularly during drought years. In contrast to earlier times, the Yellow River has never dried up since the completion of the dam. Hence, a large agricultural area benefits significantly from the water supply function of the Xiaolangdi project (Liu and Xia 2004; Ministry of Water Resources 2009).

Third, learning from lessons from the Sanmenxia Dam, Xiaolangdi Dam took the sand accumulation problem into consideration. About 60 % of the reservoir's capacity was designed to store sand coming from upstream, and it is estimated that the height of the downstream riverbed will not increase for at least 20 years. In the long-term, the dam also would control water speeds slowing the washing-out of sand along the downstream riverbed, and with collaborative efforts to restore vegetation and reduce soil erosion in upstream areas, sand accumulation and depletion downstream could reach a balance (Ministry of Water Resources 2009).

Fourth, Xiaolangdi Dam also has an electricity generation function, with an installed capacity of 1,800 MW, which could effectively meet the electricity supply for Henan Province, where it is located (Liu and Xia 2004; Ministry of Water Resources 2009).

Overall, Xiaolangdi Dam has achieved most of its goals and contributed significantly in managing the Yellow River. The sand accumulation problem has been greatly reduced, and the Yellow River is becoming more and more stable. The World Bank has praised this project, regarding it as a successful model for other projects it is supporting in developing countries (China.org.cn 2008).

Despite various achievements of the Xiaolangdi Dam, it also has had substantial social and environmental impacts. For example, 201,400 people were relocated, including 159,400 in Henan Province and 42,000 in Shanxi Province. At the maximum water level of 275 m, almost 278 km^2 of land in eight counties were submerged or indirectly affected, inundating 13,000 ha of farmland and 3,960 ha forests and orchards. Furthermore, 174 villages and 787 factories needed to be relocated, and 297 historical sites were submerged (Ministry of Water Resources 2009).

According to the requirements of the World Bank for environmental protection, the Xiaolangdi project also developed different methods to minimize its impacts on the environment. However, it still changed the landscape across a large area, and the negative effects caused by the project likely will not be known for years to come (Ministry of Water Resources 2009).

2.3.3 *The Lancang River*

2.3.3.1 A Successful Hydropower Development Model

The Upper-Mekong River, which is called the Lancang River in China, is one of Yunnan's three parallel rivers. It flows from the Tibetan Plateau, goes through Yunnan Province for 1,247 km, and runs through Myanmar, Laos, Thailand,

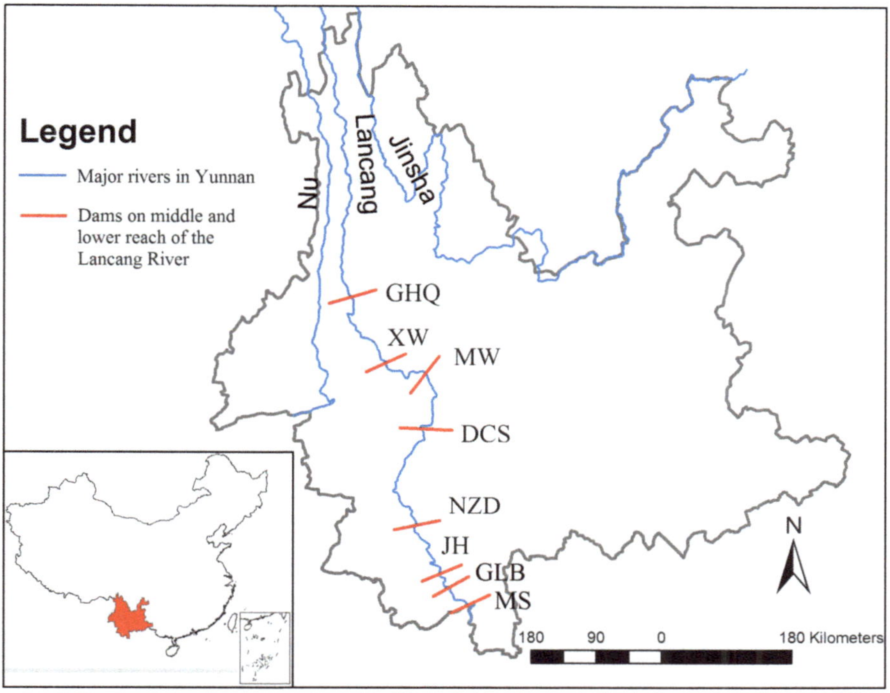

Fig. 2.2 Cascading dams on the Lancang River

Cambodia, and Vietnam before empting into South China Sea. Flowing through deep valleys, the Mekong River has an elevation drop of over 800 m in Yunnan Province, which is ideal for building dams in the view of hydrologic engineers. In the 1980s, Yunnan Province invited nationally renowned hydropower experts to design the Lancang cascading hydropower exploitation plan. These experts designed 14 cascade dams on the reach of the Lancang River in Yunnan Province, with eight dams along its middle and lower reaches. From upstream to downstream, these eight dams are: Gongguoqiao (GGQ), Xiaowan (XW), Manwan (MW), Dachaoshan (DCS), Nuozhadu (NZD), Jinghong (JH), Ganlanba (GLB), and Mengsong (MS) (Fig. 2.2) (Han 2006).

The development process of the Lancang cascade dams is a good illustration of the Chinese hydropower model, which is "watershed, cascade, rolling, comprehensive." "Watershed" means a watershed should have a whole and consistent exploitation plan, instead of separate plans for different reaches; "cascade" means for each large river, there should be a series of dams being built to take advantage of the entire river's hydropower potential; "rolling" is the strategy to use a small amount of capital to build the first of the cascade dams, then use the profit from the first dam to build the second one, then the third, and so on; "comprehensive" has two basic meanings, the first is that all the dams should regulate the flow under a comprehensive plan for the whole river, and the second is to build dams not only to generate electricity, but also as means for enhancing irrigation, water supply, navigation, and tourism (Han 2006; Duan 2009).

The first dam built on the mainstream of the Lancang River was Manwan Dam (MW, Fig. 2.2). It was completed in 1995, with an installed capacity of 1,250 MW. Manwan is the first dam in Yunnan Province that takes part in the "Transport western electricity to the east" strategy. Part of the electricity from it is transported to highly developed Guangdong Province to meet its massive energy demand. The second stage of the Manwan project was completed in May 2007, adding another 300 MW installed capacity to the hydropower station.

Dachaoshan (DCS, Fig. 2.2) was the second dam built on the Lancang River, which was completed in 2003, with an installed capacity of 1,350 MW. Its financing and management was a revolution in China's large hydropower dam construction history, which for the first time introduced the modern enterprise system into the hydropower sector.

The third dam, Xiaowan (XW, Fig. 2.2), is 70 km upstream of Manwan Dam (MW). It was completed in 2009, with the installed capacity of 4,200 MW. It is one of the two key water-control dams in the cascade dam system, having the ability to regulate flow for all the downstream dams in a long term. The height of the dam is 300 m, forming a reservoir of 14,560 million m^3.

Nuozhadu Dam (NZD, Fig. 2.2) is the largest on the Lancang River, and it is the other key water-control dam, regulating flow for three dams downstream of it, Jinghong (JH), Ganlanba (GLB), and Mengsong (MS). Nuozhadu Dam is still under construction, and the planned installed capacity is 5,850 MW. According to the official website, the dam is planned to be completed in 2017 (Xinhua News Agency 2002).

Jinghong Dam (JH, Fig. 2.2) was completed in 2009, with an installed capacity of 1,500 MW. According to an agreement between China and Thailand in 2001, a Thai company was to hold 70 % of the company's shares, and all the electricity generated from plant was to be sold to Thailand. However, to accelerate the implementation of the "transport western electricity to the east" strategy in China, it was finally decided that a Chinese company would hold all of the shares and that the electricity from Jinghong would be sent to Guangdong Province.

The other three dams, Gongguoqiao, Ganlanba, and Mengsong (GGQ, GLB, MS, Fig. 2.2), are either under construction or being designed. These dams are much smaller than the other dams, and they will not have the ability to regulate flow. The designed installed capacities for each of them are: Gongguoqiao, 750 MW; Ganlanba, 250 MW; and Mengsong, 600 MW (Xinhua News Agency 2002; Duan 2009).

2.3.3.2 Environmental and Social Consequences

Cascading dams on the Lancang will dramatically change the profile of the river, and alter its natural flow significantly (Fig. 2.3). It is also clear that Xiaowan and Nuozhadu are much larger dams than the others, and these two dams actually have

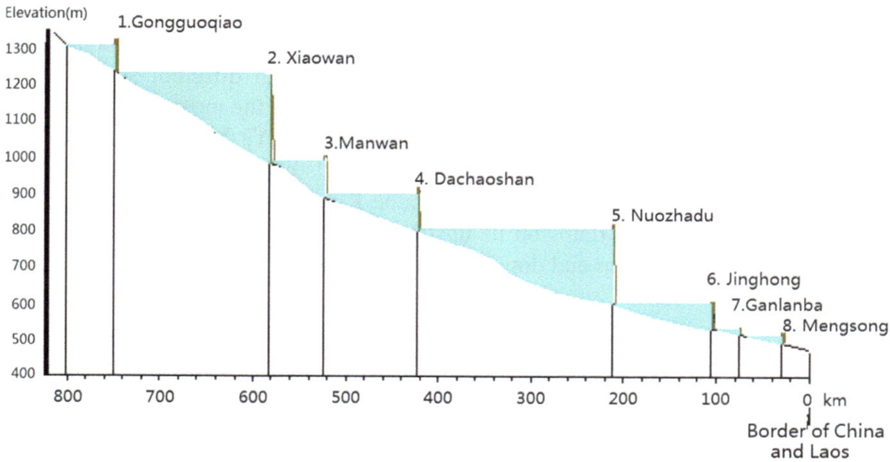

Fig. 2.3 Vertical profile of the Lancang River after dam construction (Synthesized from data on Magee 2006)

the ability to regulate the flow of the downstream dams. Such a significant change in the landscape will undoubtedly have impacts on the environment and local human communities.

The first environmental impact dams have is on local agriculture. Only 7 % of land in Yunnan is arable, and most of the arable land is along riversides. The newly formed reservoirs will submerge a large area of productive farmland and local villages will be forced to move. The people so relocated would cause secondary environmental impacts, such as the clearing of new land for agriculture and increasing deforestation and soil erosion in that area. The newly cleared farmland is generally less fertile than the original land, and therefore it will require more fertilizer, being more costly and possibly promoting more pollution of water bodies. The second impact is on fish and fishery in the Lancang River. The lower reach of the Mekong River, which is downstream to the Lancang River, is one of the largest freshwater fishing grounds in the world. It is estimated that freshwater fishes constitute 80 % of the animal protein consumption in Cambodia. During the wet season, a large area in Cambodia is inundated by the Mekong River, which provides an opportunity for many fish species to spawn (Dore and Yu 2004). Regulating the flow of the Lancang River will reduce the magnitude and frequency of floods downstream and substantially reduce fish populations. Migratory fish cannot go to upstream due to the fragmentation of the watercourse, which might cause the extinction of some species. In addition, the cascade dams also aggravate sedimentation in the Lancang River. Evidence shows that the rate of soil erosion along the Lancang River has increased in recent years, which could significantly shorten the lifespan of the dams.

Social impacts of the cascade dams include changes in local livelihoods, social inequity problems, cultural diversity loss, and international conflicts over water use.

Local farmers suffer wealth loss in both material and nonmaterial forms, such as submerged farmland and houses, being unable to use certain skills, and social capital loss. Relocation could also cause social inequities at different scales, from household to regional levels. The Lancang region is one of the most cultural diverse areas in China, with more than 20 ethnic groups (Dore and Yu 2004). A mass relocation program could cause significant cultural impacts and loss in traditional customs. The Mekong River is called "the Danube River in the East," going through seven countries. Building dams on its upstream reaches likely would heighten the conflicts between upstream and downstream residents.

2.3.4 The Jinsha River

The Jinsha River is the upper reach of Yangtze River, with a watershed of 473,200 km^2, which is about 26 % of the total Yangtze watershed. The upper reach of the Jinsha River is from Yushu to Shigu, a length of 958 km, and an elevation drop of 1,677 m; the middle reach is from Shigu to Panzhihua, which is 1,326 km in length with an elevation drop of 1,570 m; and the lower reach is from Panzhihua to Yibin, which is 782 km and has a 729 m elevation drop. The Jinsha River has the hydroelectric potential of 112,400 MW, which is about 16.7 % of China's hydroelectric potential. Most of the river runs through deep and narrow gorges, geologically suitable for building large dams (China Three Gorges Corporation 2008; Li 2009).

The planning of hydroelectric projects on the Jinsha River dates back to the 1950s. In the 1990s, according to the national hydroelectric planning strategy, 14 dams were proposed along the river. Two of these dams were to be on the upper reach, Rimian and Tuoding; eight dams were proposed for the middle reach, Hutiaoxia, Liangjiaren, Liyuan, Ahai, Jinanqiao, Longkaikou, Ludila, and Guanyinyan; and four dams were identified for the lower reach, Wudongde, Baihetan, Xiluodu, and Xiangjiaba. The State Development and Reform Commission dictated the order of their construction: dams on lower reach were to be built first, followed by dams on the middle reach, and last those on the upper reach (Li 2009).

In 2002, the China Three Gorges Corporation, a state-owned company, obtained the right to build the four dams on the lower reach. The Xiluodu Dam project commenced in December 2005, and Xiangjiaba began in December the following year. The other two lower reach dams, Wudongde and Baihetan, are presently going through feasibility studies and environmental impact assessments. The proposed total installed capacity of the four dams is twice as large as that of the Three Gorges Dam (Li 2009).

At the end of 2005, a new state-owned company, the Yunnan Jinsha River Hydropower Corporation, was founded, with the goal of building the eight middle reach dams. According to the *Jinsha Middle Reach Hydropower Planning Report* in 2003, the total installed capacity of these dams will be 20,580 MW. The shareholders in the Corporation are actually several big electricity companies in China: Huadian Corporation, Huaneng Corporation, Datang Corporation, Hanneng

Corporation, and Yunnan Investment Corporation. Each of these companies is in charge of one or more dams; for example, Huadian is responsible for building Ludila Dam, and Huaneng is responsible for Longkaikou Dam. The fact that different companies manage different dams on the middle reach of the Jinsha River might make flow regulation more complex and politicized. Actually, flow regulation of the entire Yangtze River lacks a comprehensive management strategy, which impairs the functioning of its dams for flood control and supplying water. For example, when the Yangtze watershed suffers from drought, downstream areas need more water for irrigation and urban consumption, but in many cases the upstream dams hold the water to generate electricity, which aggravates drought problems downstream. When there is heavy precipitation in the watershed, the upstream dams generally discharge their stored water to release pressure on the dams, which makes the situation of the already flooded downstream areas even worse (Li 2009; South China Weekends 2010).

Similar to the situations of many large dam projects on other rivers, there have been heated debates about the dams on the Jinsha River in recent years. Because of the massive profits arising from the production of hydroelectricity, hydropower companies and local governments eagerly push the dam construction process forward. But the Ministry of Environmental Protection and environmental NGOs also try their best to prevent some of the construction (South China Weekends 2010; Li 2009).

Opponents of the dam projects on not only the Jinsha River, but also other major rivers in Southwestern China, include some Chinese NGOs, journalists, scientists, as well as some environmental protection agencies. Their major arguments are summarized below.

1. The Southwestern Chinese Rivers are marked by high biodiversity and the "Three Parallel Rivers" are identified as part of the World Natural Heritage by UNESCO. The dam projects might cause irreversible impacts on the river gorges ecosystems and pristine scenery.
2. These rivers run across a region with very high ethnic and cultural diversity, and dam construction would require relocating thousands of these indigenous people, which would damage their unique customs and cultures.
3. Some experiences with dam-caused relocation programs in Yunnan Province showed that they significantly lowered the standards of living of the relocated people. Therefore, it is socially unfair for local people to sacrifice their livelihoods while hydroelectric companies reap huge benefits from the project.

The proponents of the dams care more about local development, and they think the dam projects will provide golden opportunities. Those people include local government officials, managers of state-owned hydroelectricity companies, and some scientists. Their major points are summarized below.

1. Because of population growth and extreme poverty, local people have already destroyed the primary forest below the elevation of 2,000 m. Therefore, dam construction would only cause inundation of these already ecologically destroyed areas and would not have substantial impacts on primary ecosystems above that elevation.

Actually, dams might even benefit the ecosystem, since the electricity generated could reduce the mass consumption of fuel wood and save remaining intact forests.
2. Hydroelectric projects could bring prime development opportunities to this remote area. The dam projects could generate substantial financial returns, and, even though only a small portion would be retained locally, this would result in at least a tenfold increase in the local revenue. This is why the local government is among the most active advocators of the dam project.
3. It is undeniable that the dam projects will have significant impacts on the river species, particularly some migratory fishes. But the environmental impact assessment showed that among many indigenous fish species, only a couple of them would likely to be driven to extinction. Hence, the value of one species cannot be compared with the development opportunities for millions of people.
4. Population density in the proposed project area is relatively low compared to other hydraulic project areas. For example, even though the installed capacity is larger than Three Gorges Dam, the dams would require the relocation of less than 10 % of the people displaced by the Three Gorges Dam project. Therefore, the social impacts of the dam project are also relatively low.

Unlike the Nu River situation discussed in Chap. 4, where environmental NGOs successfully halted the proposed dam construction, there are several dams already under construction along the Jinsha River. The fragmentation of the Jinsha River has extensive social and environmental impacts. Unfortunately, there appear to be fewer published studies addressing the potential impacts of dams on the Jinsha River compared to other case studies presented in this section (South China Weekends 2010; Li 2009).

References

Chen C-TA (2002) The impact of dams on fisheries: case of the Three Gorges Dam. In: Steffen W, Jaeger J, Carson DJ, Bradshaw C (eds) Challenges of a changing earth. Springer, Berlin

Childs-Johnson E, Sullivan LR (1996) The Three Gorges Dam and the fate of China's southern heritage. Orientations 27:55–61

China Hydropower Engineering Web (2011) http://www.slsdgc.com.cn/show.php?contentid=7743

China Three Gorges Corporation (2008) Special issue on Jinsha River hydropower development. http://www.ctgpc.com.cn/jsjsdkf/special_news.php

China.Org.CN (2008) The Xiaolangdi Key Water-control Project on the Yellow River. http://www.china.com.cn/aboutchina/data/zmslgc/2008-05/08/content_15122872.htm

Dore J, Yu X (2004) Yunnan hydropower expansion: update on China's energy industry reforms and the Nu, Lancang and Jinsha hydropower dams

Duan X (2009) Experiences and recommendations for hydropower development on Lancang River. http://www.wcb.yn.gov.cn/slsd/ztyj/3857.html

Han, D (2006) Expedite the Hydropower Development on Lancang River. http://www.chinapower.com.cn/article/1043/art1043147.asp

Henan Xinhua News Agency (1997) Xiaolangdi project will not become the second Sanmenxia Dam. http://www.ha.xinhuanet.com/xhzt/2006-10/24/content_8336708.htm

Jackson S, Sleigh A (2000) Resettlement for China's Three Gorges Dam: socio-economic impact and institutional tensions. Comm Post-Commun Stud 33:223–241
Jackson S, Sleigh AC (2001) The political economy and socio-economic impact of China's Three Gorges Dam. Asian Stud Rev 25:57–72
Li W (2009) Halt of two hydropower dams on Jinsha River reflects the dilemma of hydropower development in China. Life Week (in Chinese)
Li H, Waley P, Rees P (2001) Reservoir resettlement in China: past experience and the Three Gorges Dam. Geogr J 167:195–212
Liu C, Xia J (2004) Water problems and hydrological research in the Yellow River and the Huai and Hai River basins of China. Hydrol Process 18:2197–2210
Magee DL (2006) New energy geographics: powershed politics and hydropower decision making in Yunnan, China. PhD Dissertation, University of Washington
Ministry of Water Resources (2009) Yellow River Water and Hydropower Development Corporation, Ministry of Water Resources. http://www.xiaolangdi.com.cn/xiaolangdi/
Muller B, Berg M, Yao ZP, Zhang XF, Wang D, Pfluger A (2008) How polluted is the Yangtze river? Water quality downstream from the Three Gorges Dam. Sci Total Environ 402:232–247
Pan J, He J (2000) Large dams in China: a fifty-year review, China. WaterPower Press, Beijing
People's Daily Online (2010) Three Gorges Dam relocated people married local girl. People's Daily Online. http://energy.people.com.cn/GB/12773812.html
South China Weekends (2010) Two hydropower dam projects resume one year after rejection. http://www.infzm.com/content/50594
Stone R (2008) Three Gorges Dam: into the unknown. Science 321:628–632
Tan Y, Liu H (2003) Sanmenxia Dam: the dangerous status of the former "first dam in China." http://www.people.com.cn/GB/huanbao/1072/2206736.html. Accessed May 15, 2012
Tianjin E-North Netnews (2003) A 50 year reflection on The Sanmenxia Dam. http://news.enorth.com.cn/system/2003/12/03/000678411.shtml. Accessed May 15, 2012
WCD (2000) Dams and development: a new framework for decision-making. World Commission on Dams
Wu J, Huang J, Han X, Gao X, He F, Jiang M, Jiang Z, Primack RB, Shen Z (2004) The three Gorges dam: an ecological perspective. Front Ecol Environ 2:241–248
Xinhua News Agency (2002) The history of lancang river hydropower development. http://news.xinhuanet.com/fortune/2002-01/20/content_245846.htm
Xinhua News Agency (2003a) Background of the Three Gorges Project. http://news.xinhuanet.com/ziliao/2003-05/30/content_894678.htm
Xinhua News Agency (2003b) Background of the Three Gorges Project. http://news.xinhuanet.com/ziliao/2003-05/30/content_894678.htm. Accessed May 15, 2012

Chapter 3
Environmental Impacts of Dams in China: Focusing on Biological Diversity and Ecological Integrity

3.1 An Overview of Environmental Problems Caused by Dams in China

With the growing awareness concerning the environmental impacts of coal-fired power plants, China has shifted to exploiting hydropower, a cleaner energy resource. China has become the world's largest hydropower producer having built some 87,000 dams and reservoirs, about one-third of which are for hydroelectricity production. According to an estimate by the Chinese Government, the Three Gorge Dam can generate the equivalent amount of electricity as burning 50 million tons of coal annually, thus reducing greenhouse gas emissions by100 million tons of carbon dioxide, 1.2–2 million tons of sulfur dioxide, and 10,000 tons of carbon monoxide as well as eliminating the release of large quantities of particulate matter (Xinhua News Agency 2007). However, many argue that hydroelectric power should not be considered a renewable energy source because of the irreversible environmental damages caused by these projects. The environmental debate concerning the construction of large dam is not new in China and has been ongoing for decades since the first large Soviet-style dam, Sanmenxia Dam, was built in 1957 on the main channel of the Yellow River. As Shapiro (2001) states about the Chinese hydraulic engineer, Huang Wanli, who fought against dams, "Many years of his own practical research conducted on the Yellow and other major rivers had shaped his conviction that damming the main streams of large waterways ran contrary to the laws of nature."

The environmental impacts caused by large dams in China are no different than those that occur in other parts of the world (see Sect. 1.3). China also has some additional environmental concerns, such as an increased danger of landslides and earthquakes in areas of high seismicity, such as Western China. More than 130 large dams are being built in western China with 98.6 % of these located in moderate to very high seismic hazard zones (Probe International 2012). Here, we synthesize the key environmental impacts caused by large dams constructed along eight major

P. Wang et al., *The Large Dam Dilemma: An Exploration of the Impacts of Hydro Projects on People and the Environment in China*, DOI 10.1007/978-94-007-7630-2_3,

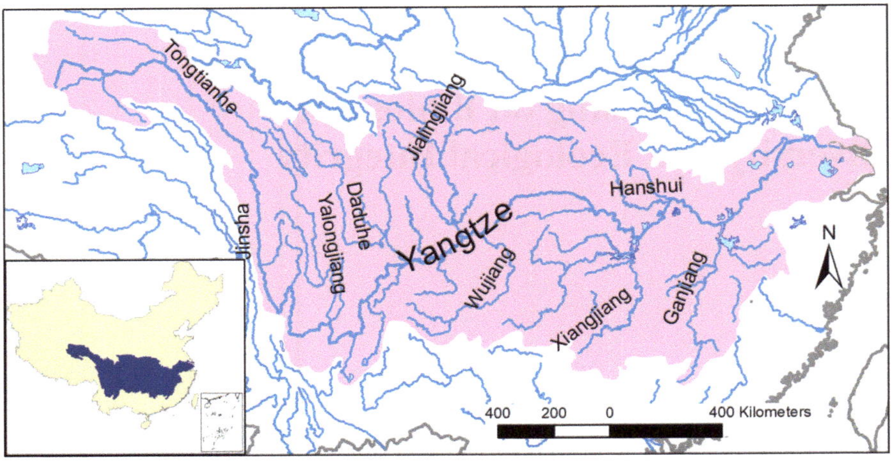

Fig. 3.1 Map of the Yangtze River Basin

river basins in China, with particular focus on ecological impacts. In the following Sect. 3.2, we specifically discuss the effects of large dams on biological diversity and ecological integrity to support a case study presented in Sect. 3.3.

3.1.1 The Yangtze River Basin

The Yangtze (also called Chang Jiang) is the longest river in China and all of Asia, and the third longest in the world. It originates in the glaciers on the Qinghai-Tibetan Plateau and flows 6,418 km south and then eastward across southwest, central, and eastern China before emptying into the East China Sea at Shanghai, one of China's largest cities (Fig. 3.1). The basin is home to 400 million people, making it one of the most densely populated river basins anywhere on earth—with an average population of 214 people per km^2 (WRI et al. 2003). The Yangtze River Basin is already highly fragmented by a significant number of dams over 15 m high, including one of the world's largest, the Three Gorges Dam. The additional 46 dams planned or under construction will put further pressure on an already stressed basin (WWF 2004).

Alteration of the flow regime is one of the biggest environmental issues associated with dam construction along the Yangtze River. Building of the Three Gorges Dam has been widely criticized since the most severe drought in Yangtze River Basin in the history was recorded in 2011 (Lu et al. 2011), although the debate of "to build or not to build" has lasted for decades (Dong 2005). Soon after the droughts, torrential rains triggered floods and landslides causing massive economic loss and environmental damage (Xinhua News Agency 2011). While it was believed that the Three Gorges Dam was built to control floods, generate clean energy, and improve irrigation capacity, the severe downstream flood and drought have put a question mark on its flood control objectives (Ran et al. 2012). The Chinese Government conceded that there were some problems associated with the Three Gorges Dam project in their statement, "Although the Three Gorges Dam project provides huge comprehensive benefits, urgent problems must be resolved regarding the smooth relocation of residents, ecological protection and geological disaster prevention" (New York Times 2011).

Siltation in the reservoir is another problem associated with dams in the Yangtze River Basin. For example, an estimated 530 million tons of silt already has accumulated in the reservoir of the Three Gorges Dam due to the reduced flow velocity behind the dam (Mount Holyoke College 2007). The rising silt levels could potentially block the sluice gates that are essential for controlling water levels behind the dam, which could cause more flooding to occur upstream in the event of a heavy rainfall (Mount Holyoke College 2007). Also, reduced downstream sediment flows may negatively impact downstream floodplains, such as the Jianghan (Yangtze River and Hanjiang River) Plain and Dongting Lake Plain (Li et al. 2000), threatening these floodplain areas.

The reservoir acts as a trap for debris, pollutants, and toxins being carried by the river contaminating the water (Mount Holyoke College 2007). Approximately ten million tons of plastic bags, bottles, animal corpses, trees, and other detritus that would have otherwise flowed downstream and out to sea have accumulated behind the Three Gorges Dam. The reservoir itself flooded 1,600 abandoned factories, mines, dumps, and potential toxic waste sites. More than 265 billion gallons of raw sewage are dumped into the Yangtze River leading to frequent algal blooms and eutrophication throughout the basin (People's Daily 2007). The untreated wastes, together with other nonpoint pollutants from agriculture and point pollutants from industries (Sichuan Daily, 2004), have already made the Yangtze River one of the most polluted rivers in the world (Fig. 3.2). In addition, the decomposition of vegetation and organic materials accumulated in the dam's reservoir is source of greenhouse gases. Although the methane emission rate in the Three Gorge reservoir is relatively low (0.26 ± 0.38 mg CH4 m^2/h) compared with other hydropower reservoirs (Chen et al. 2011), the total amount released is high due to the large surface area of the reservoir.

Increased risk of landslides and earthquakes are current environmental concerns associated with the Three Gorges Dam. Since construction of the dam, a total of 36 km of shoreline in 91 locations has collapsed with some landslides triggering 50 m-high waves on its reservoir (Mount Holyoke College 2007). In Fengjie County, upstream of the Three Gorges Dam, there are more than 800 disaster-prone areas and the potential for geological catastrophe is threatening the lives of millions of residents (Mount Holyoke College 2007). Additionally, in the upper Yangtze River Basin, an area of high seismicity in Western China, reservoir-induced seismicity is likely to increase the frequency and perhaps magnitude of earthquakes (Probe International 2012). In the event of an earthquake or dam collapse, over 360 million people who live within the watershed of the Yangtze River will be in danger (Mount Holyoke College 2007). For example, the Zipingpu Dam built in a moderate seismic zone in the upper Yangtze River Basin is now thought to have triggered the

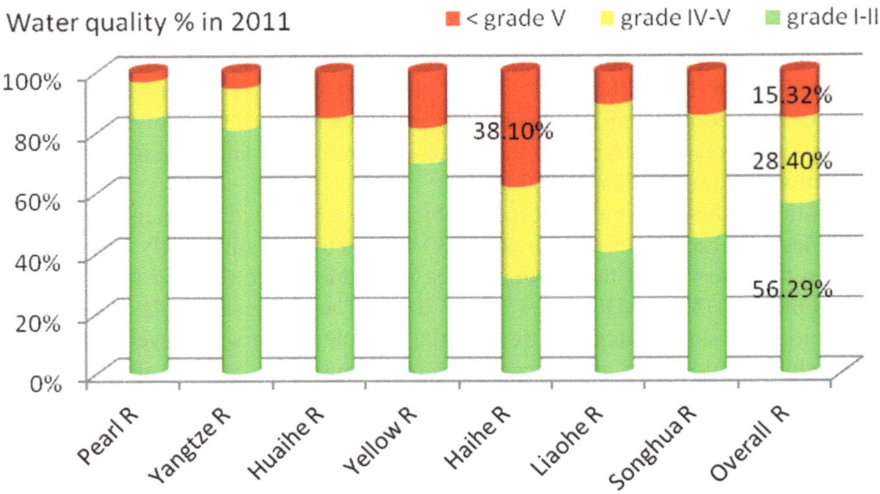

Fig. 3.2 Fresh water quality in China: an overall grim situation of water quality in major rivers in 2011 (Adopted from http://asianfootprint.blogspot.com/2012/11/fresh-water-environment-in-china.html). *Note*: China uses a six-grade classification scheme for water quality. Grade I and II are the best. Water no worse than grade III can be used for drinking. Grade V can be used for irrigation. Water less than grade V cannot be used for irrigation

magnitude 7.9 Sichuan earthquake in 2008 that killed an estimated 80,000 people (Mount Holyoke College 2007).

The Yangtze River Basin has a biologically rich terrestrial and freshwater flora and fauna that originated from tropical, subtropical, and temperate zones (Wu et al. 2004). Environmental alterations associated with large dams may result in a number of regional changes in terrestrial and aquatic biodiversity as well as ecosystem structure and functioning (Wu et al. 2004). With the completion of the Gezhouba Dam in 1981, 40 km downstream from the Three Gorges Dam site, there was a rapid and sharp decline in the populations of three of China's most famous ancient and endangered fish species, the Chinese sturgeon (*Acipenser sinensis*), the river sturgeon (*Lipotes vexillifer*), and Chinese paddlefish (*Psephurus gladius*) (Xie et al. 2003). Below the two high head dams, Xinanjiang Dam (Qiantang River) and Danjiangkou Dam (Han River), fish spawning has been delayed for 20–60 days due to lower water temperatures (Zhong and Power 1996). Changes in the flow regime resulted in the extinction of *Macrura reevesii*, a highly valued fish, in the Qiantang River. The Three Gorges Dam marked the upper limit in the distribution of the Yangtze River dolphin or Baiji dolphin (*Lipotes vexillifer*), which is the most threatened cetacean in the world (IUCN 2002). The construction of the Three Gorges Dam and other large dams on tributaries in the middle reaches of the basin are exacerbating pressures on the remaining population of these dolphins (Revenga 2003). In 2006, scientists exploring the Yangtze River failed to find a single Baiji dolphin prompting fears that mankind may have killed off its first species of cetacean (The Guardian 2007).

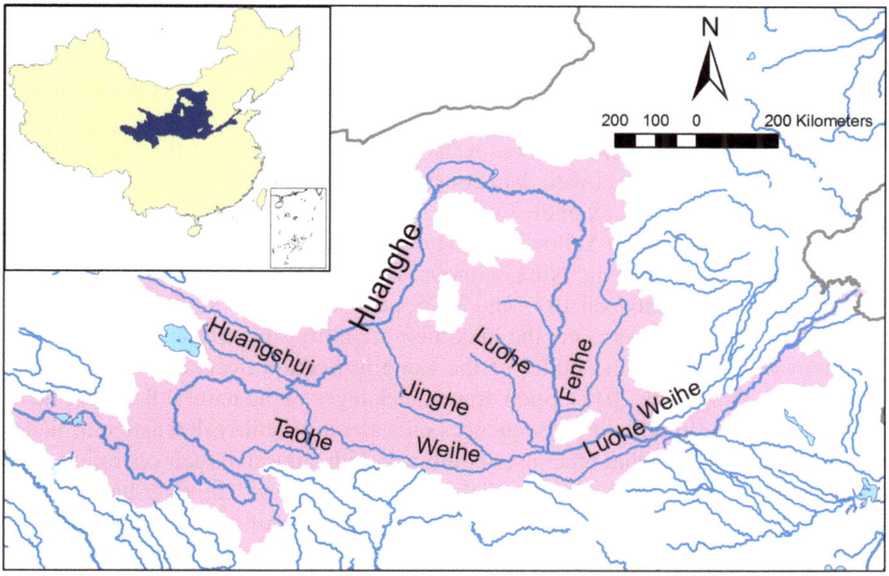

Fig. 3.3 Map of the Yellow River Basin

Waterfowl also are impacted by large dam construction in the Yangtze River Basin. For example, the rich mudflats of Poyang Lake are the major wintering sites of the Siberian crane (*Grus leucogeranus*), with 95 % of the population overwintering there. However, habitat loss in the lake associated with the construction of the Three Gorges Dam now poses a major threat to this endangered bird species (WWF 2004). In addition, dams can alter the riparian zone along a river leading to the destruction of riparian vegetation and degradation of waterfowl habitats. As a consequence, riparian biodiversity can be lost. For example, 400–770 vascular upstream plant species will be flooded when the Three Gorges Dam reaches its planned maximum water level (Wu et al. 2004). Two plant species, *Myricaria laxiflora* and *Adiantum reniforme* var. *sinese* likely became extinct due to inundation (Xie et al. 2006); two other species, *Securinega wuxiensis* and *Neyraudia wushanica*, lost critical habitat (Huang 2001).

3.1.2 *The Yellow River Basin*

The Yellow River (also called Huang He) originates in the Bayan Har Mountains in Qinghai Province in western China and is the second longest river in China and the sixth longest in the world (Fig. 3.3). It flows through nine provinces and empties into the Bohai Sea, with an estimated length of 5,464 km. The Yellow River is called "China's Mother River" or "the Cradle of Chinese Civilization" as its basin was the birthplace of ancient Chinese civilizations and the most prosperous region in early Chinese history. This river already is heavily dammed and an additional eight large dams are planned or under construction, which will put further pressures on the basin (WWF 2004).

Like the Yangtze River, alteration of the flow regime is one of the biggest environmental issues associated with dam construction along the Yellow River. Ouyang and colleagues' (2011) study of the Longliu section (from Longyangxia to Liujiaxia) of the upper reaches of the Yellow River showed that after the construction of eight cascade dams (Longyangxia, Nina, Lijiaxia, Zhiganglaka, Kongyang, Gongboxia, Suzhi, and Liujiaxi) between 1977 and 2006, peak streamflow shifted from June to May and October. Furthermore, the maximum monthly streamflow velocity difference between the inlet and outlet of the Longliu Section decreased from 430 to 115 m^3/s (Ouyang et al. 2011). Such dramatic changes in the natural flow regime of the river can have important consequences on water availability downstream. Water diversion for irrigation and other uses combined with dams and their operation have resulted in the lower reaches of the Yellow River drying up. The river has dried up 30 times since 1972 with the longest period of no-flow lasting for 230 days in 1997 (He et al. 2012). Timely releases of reservoir water have prevented the river from drying up since 2000.

Sedimentation is a serious problem in the Yellow River Basin. The river gets its name from the yellow silt it picks up in the Shaanxi Loess Plateau and it has a naturally high sediment load. Because of these high sediment loads, the river is prone to flooding; silt is deposited on the riverbed causing the river levels to rise. Levees have been built along the lower 800 km of the river to contain the water, but the riverbed continues to rise and the river is now elevated above the surrounding landscape (Wang et al. 2007b). The levees along the "suspended river" have to be continuously raised and consolidated to constrain water flow within the channel (Ran et al. 2012), and any breach in a levee will result in massive flooding. The first major water control project on the Yellow River was the Sanmenxia Dam constructed in 1957–1960 in the middle reaches of the Yellow River. As discussed earlier, the dam was poorly designed and did not account for the high sediment loads carried by the river, and the reservoir silted up 2 years after it began operation causing floods in upstream areas and threatening river navigation (Wang et al. 2005). To mitigate the negative impacts of sedimentation on reservoir storage capacity and downstream flow regimes, the dam has been reconstructed to provide high sediment releasing capacity, and dam operation has achieved a balance between sediment inflow and outflow (Wang et al. 2005). Subsequently, the Xiaolangdi Dam, China's second largest dam, was built downstream of the Sanmenxia Dam to control flooding by slowing the dangerous build-up of silt on the lower reaches of the river (Hays 2009).

Dams are having a negative impact on the water quality of the Yellow River Basin as pollutants accumulate behind dams instead of being flushed downstream (Hays 2009). The Yellow River travels through major industrial areas including thousands of petrochemical plants, large population centers, and China's major coal producing area. Over 30 % of water samples taken from the Yellow River in various places registered worse than Level 5, meaning the water was unfit for drinking, agriculture, or industrial use (Fig. 3.2). Four billion tons of wastewater, which is 10 % of the river's volume, flows annually into the Yellow River (Hays 2009), making it the most polluted river in the world.

Reduced water flow, severe pollution, dams blocking fish migration routes, and overfishing are threatening fish diversity in the Yellow River Basin. An unnamed agriculture ministry official in China said, "The Yellow River used to be host to more than 150 species of fish, but a third of them are now extinct, including some precious ones" (The Guardian 2007). According to a survey conducted by the People's Daily, catches of fish have fallen 40 % from an annual average of 700 tones (The Guardian 2007). In addition, terrestrial plants and animals in the basin also can be impacted by land use/cover changes associated with the construction and operation of dams. Large dam projects in the Yellow River Basin threaten four critical habitats for endemic birds and one wetland of international significance according to the Ramsar Convention on Wetlands (WWF 2004). Construction of the cascade of dams in the Longliu Section (from Longyangxia to Liujiaxia) of the upper reaches resulted in the loss of large areas of grassland (Ouyang 2011).

3.1.3 Lancang River Basin

The Lancang (Upper-Mekong) River situated in southeastern Eurasia with a length of 4,909 km originates as Guyong-Pudigao creek near the foot of Mt. Jifu on the Qinghai-Tibet Plateau and discharges into the South China Sea (Liu et al. 2007). It is a famous transboundary river and is commonly divided into two parts: the Upper-Mekong River Basin, including portions in China (called Lancang River in Chinese) and Myanmar; and the Lower-Mekong Basin, including portions in Laos, Thailand, Cambodia, and Vietnam (MRC 2010). This river remains mostly undammed on the mainstream, except for eight cascading hydropower dams constructed along the Upper-Mekong River in China (Nilsson et al. 2005; Grumbine and Xu 2011) (Fig. 3.4). Another 11 dams have been proposed on the Lower-Mekong River in Laos as of September of 2010 (Grumbine and Xu 2011).

Alteration of the flow regime has been widely documented as a major environmental impact of large dams in the Upper-Mekong River Basin. Recent research shows that the overall degree of hydrologic alteration associated with dam construction and precipitation variation is 25.2, 25.3, and 29.1 %, respectively, for the upstream, midstream, and downstream areas of the Manwan Dam (located in middle reaches of Lancang River) (Zhao et al. 2012a). He et al. (He et al. 2007) found that the Manwan and Dachaoshan dams can influence the downstream flow regime of the Mekong River, but their impacts are insignificant in scale compared to changes in natural daily flow stages and is limited geographically to the narrow channel north of Vientiane. He et al. (2007) further stressed that the water flow below the Yunjinghong hydrologic station (located in lower reaches of the Lancang River) is only 15.4 % of total flow in the Mekong River, and the primary factor changing the flow volume in the Lower-Mekong River is climate variability rather

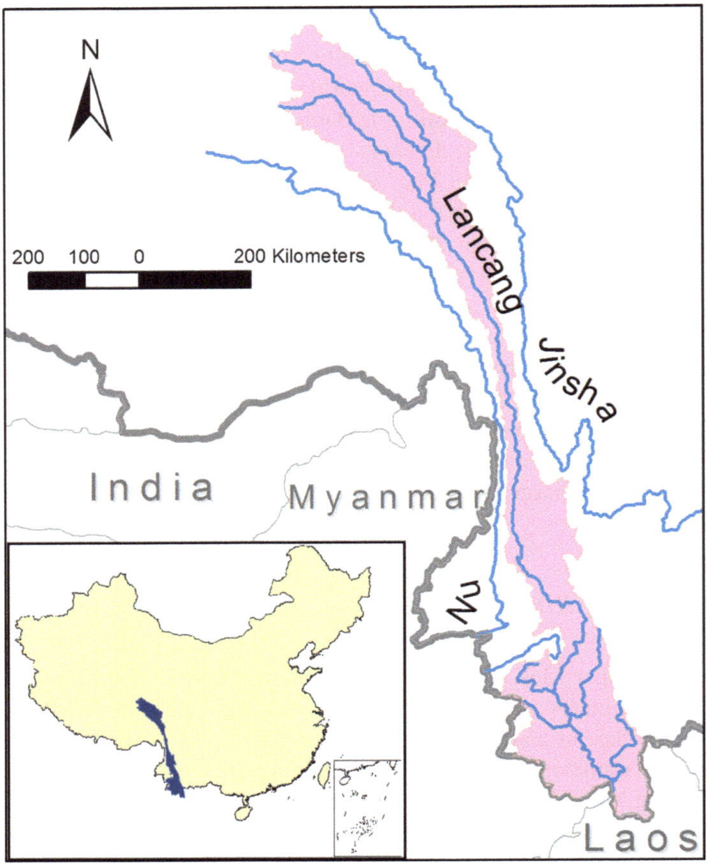

Fig. 3.4 Map of the Lancang (Upper-Mekong) River

than upstream dam construction and operation. Li et al. (2006) reported that the normal operation of the cascading dams of Manwan, Dachaoshan, and Jinghong in the upstream reaches of the Lancang River has minor interim effects on downstream monthly and annual water levels, but may have major impacts on short-term daily water levels.

Sedimentation of the dams in the Lancang River Basin has also been well studied. Fu et al. (2008) found that over the 11 years since the Manwan Dam started to store water in 1993 when the Dachaoshan Dam began to store water in 2003, the annual mean sediment storage in the Manwan Reservoir was about 26.9–28.5 × 10^6 tons, equivalent to a loss of 21.5–22.8 % of the reservoir's total storage capacity. Liu et al. (2012) reported a remarkable reduction in annual sediment load at the Gajiu hydrological station (located downstream of the Manwan Dam) after the Manwan hydropower station was put into operation in 1995, declining to over

60 % of the mean pre-dam value. Both Fu et al. (2008) and Liu et al. (2012) found there were poor correlations between upstream and downstream sedimentation at the dammed sections of the river, indicating that a large dam in the upper reaches of a river may have little effect downstream sedimentation.

Water pollution associated with the large dams in the Lancang River Basin has not been well studied. However, recent research has shown that there is the potential risk of heavy metal contamination in the sediments of the reservoir of the Manwan Dam due to the accumulation of As, Cd, Cr, Cu, Pb, and Zn in sediments that were measured at levels above the threshold established by both the United States and Canada. Moreover, the concentration of Hg in the Manwan Reservoir exceeded recommended levels of water quality standards in China, indicating that the dam may be impacting the water quality through the accumulation of heavy metals in the reservoir's water (Zhao et al. 2012c).

Habitat degradation and biodiversity loss in both aquatic and terrestrial ecosystems are another major environmental cost of constructing dams in the basin. This region is characterized as a biodiversity hotspot and a fragile ecoregion (Zhao et al. 2012c). Altered flow regimes, sedimentation, and pollution will degrade the feeding and breeding habitats of fish along the Mekong River (Kummu and Sarkkula 2008; Kang et al. 2009). Moreover, the dams are physical barriers for fish migrations. As many as 40–70 % of the fish species in the Mekong River migrate long distances along the mainstream and into its tributaries (Barlow et al. 2008; Baran and Myschowoda 2009; Kang et al. 2009). Although the impacts of dams on the upper reaches (Lancang River) of the river basin and on individual tributaries will be restricted to the fish populations that use these reaches, these populations contribute substantially to fish production along large stretches of the river (Poulsen et al. 2002). Dugan and colleague's (2010) review showed that current moves towards dam construction on the Mekong River will not stop; therefore, innovative technology and conservation strategies are needed to reduce the loss of fish diversity and ecosystem services provided by the river. For example, two tributaries of the upper reaches of the Lancang River, the Buyuan and Nanla rivers, are important feeding and spawning habitats for upstream migrant fish species and should be protected (Kang et al. 2009).

Dam construction and operation has led to land use/cover changes, alterations in water levels, thermal stratification, and water quality deterioration, all of which may influence the structure and function of riparian vegetation (Li et al. 2012a). The series of hydropower dams can induce habitat fragmentation, reduce the distribution ranges of primary vegetation, reduce the complexity of the vegetation types along the river, and potentially lead to the loss of primary vegetation in the whole watershed (Li et al. 2012c). Two endangered shrub species, *Homonoia riparia* and *Phoenix roebelenii*, are currently threatened by the dam impoundment (Li et al. 2012a), and one aggressive exotic species with extremely strong invasion capabilities, *Eupatorium adenophorum*, has spread rapidly along the riparian corridor below the Manwan Dam when the water level drops in association with the operation of the dam (Li et al. 2012d).

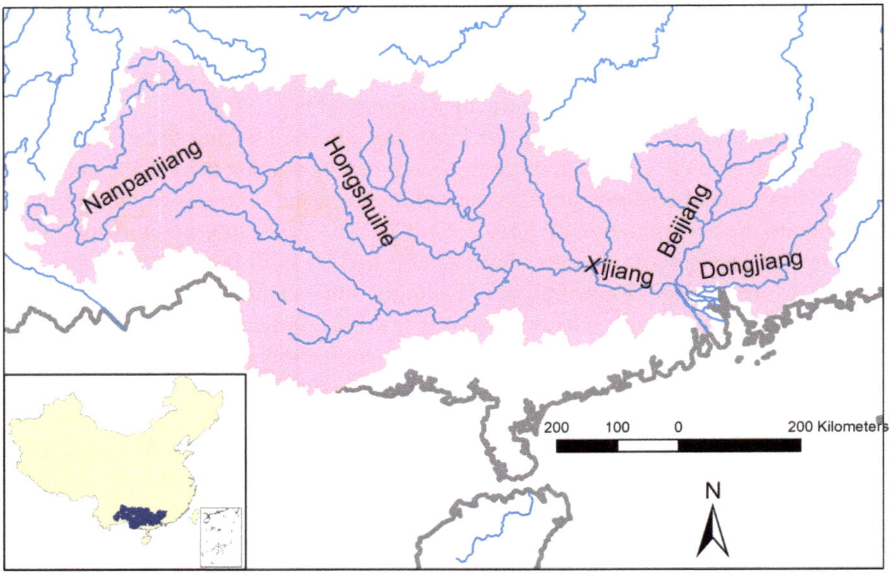

Fig. 3.5 Map of the Pearl River Basin

3.1.4 Pearl River Basin

The Pearl River (also called Zhu Jiang) is China's third longest river (2,400 km, after the Yangtze River and the Yellow River), and second largest by volume (after the Yangtze). It is the largest river in southern China, flowing to the South China Sea between Hong Kong and Macao (Fig. 3.5). The river basin drains the majority of south central China, as well as parts of southwestern China and northeastern Vietnam (The Ministry of Water Resources 2013). There are a large number of existing dams and another ten are under construction (WWF 2004).

Similar to other large dam projects in the major river basins of China, the dams in the Pearl River Basin have similar environmental impacts of flow regime alteration, sedimentation, pollution, ecosystem degradation, and loss of biodiversity. Since the 1960s, increased dam building and water extraction through diking have changed land use and land cover affecting the surface runoff pattern in the basin with a gradual rise in the water table, especially in the estuary areas (Weng 2007). In addition, unregulated sediment mining associated with dam building and dike linking has changed the longitudinal profiles of the river channels affecting flood discharge and causing the penetration of saline water (Qian 2005; Wang et al. 2005). The third largest dam in China, the Longtan Dam, was completed in 2009 on the Hongshui River, a tributary of the Pearl River. Siltation is a concern and local officials stated "deforestation in the area has caused severe erosion and power stations downstream are already facing serious problems because of silting" (BBC 2001).

Although the impacts of dams on the water quality of the Pearl River Basin are poorly known, the river suffers from serious water pollution. According to the National Marine Environmental Monitoring Centre, the Pearl River dumps 8,655 tons of heavy metals, 65,637 tons of nitrates and ammonia, and 59,853 tons of petrol into the sea each year (AsiaNews 2005). As reported by China Daily (2011), China's Fishery Eco-environment Report tallied 165 pollution incidents in the Pearl River in 2009, leading to economic losses totaling 56 million yuan (US$8.6 million). Undoubtedly, as seen elsewhere, pollutants can accumulate in the sediments and water behind large dams. As a consequence, the Pearl River is becoming one of the most polluted waterways in China.

The basin has a rich amphibian diversity with 127 recorded species, and a large bird population with four endemic bird areas that host migratory species which traverse the East Asian Australasian flyway (WRI et al. 2003). There is concern that these populations might be negatively impacted by the dams. In addition, two endemic fish species, the Chinese gizzard shad (*Clupanodon thrissa*) and Reeve's shad (*Tenualosa reevesii*), have disappeared altogether from the regulated parts of the river, and the populations of major carp species (such as *Cirrhinus molitorella*) no longer sustain a viable fisheries industry. A China Daily (2011) interview reported that, "Anadromous fish, which live in the sea but swim upriver to reproduce, have been forced to spawn before arriving at the best sites ... because traditional grounds are blocked by hydro projects; as a result, their young have slimmer chances of surviving, largely because their migration routes back to the ocean are not long enough to allow them to mature." It was also noted that "there have been no recorded catches of Chinese sturgeon in recent years, although about 400,000 kg of the sturgeon were produced each year in the 1930s in the Xi Jiang alone" (China Daily 2011).

3.1.5 Huai River Basin

The Huai River originates in the Tongbai Mountains in Henan Province and is located about midway between the Yellow and Yangtze Rivers. It flows through southern Henan, northern Anhui, and northern Jiangsu, entering the Yangtze River at Jiangdu, Yangzhou (Fig. 3.6). The river has a length of 1,078 km and a drainage area of 174,000 km^2 (The Ministry of Water Resources 2013). As the river does not flow all the way to the sea, it is notoriously vulnerable to flooding. Therefore, many dams, including 16 large dams, have been built on the river primarily for flood control, but also for irrigation and power generation (Jiang et al. 2009).

Similar to other large dams in major rivers worldwide (Liu and Xia 2004; Stone and Jia 2006; Wang et al. 2006), the large dams in the Huai River Basin have greatly changed the flow regime, water–sediment properties, and habitats of organisms living in or along the river, which has affected the natural ecosystem and environment (Jiang et al. 2009). The most serious environmental problem in the Huai River Basin is water pollution causing massive social, ecological, and economic losses. As reported by the People's Daily (2006), "almost all the tributaries of the Huaihe

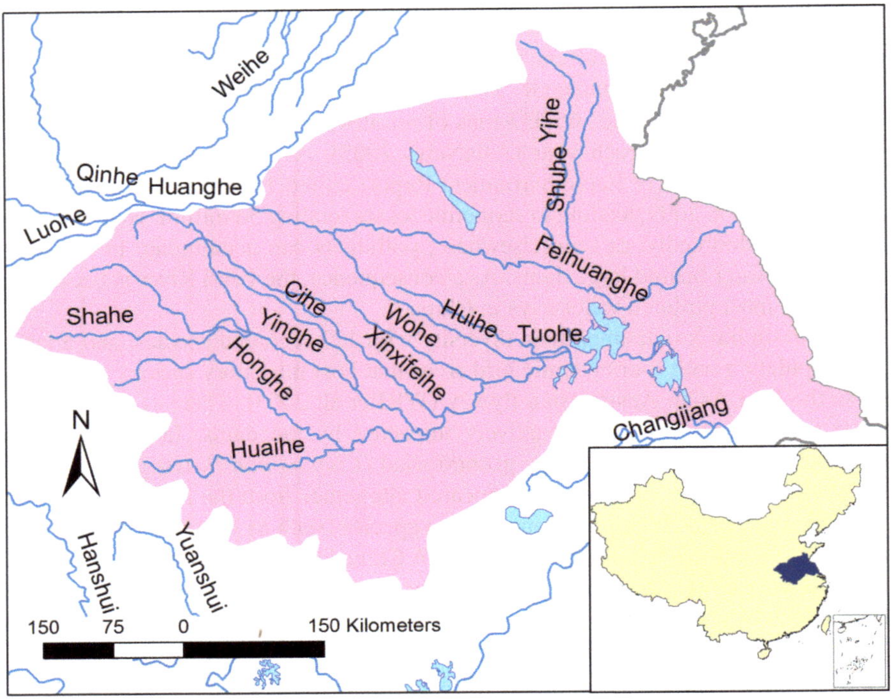

Fig. 3.6 Map of the Huai River Basin

River don't have any fish or anything alive and 70 % of the water quality is under class V, very poisonous and can not be used for any purpose." Although many scholars assume these environmental disasters to be highly related to the large dam construction and operation, a simulation modeling conducted by Jiang and colleagues (2009) showed that the increasing discharge at Hengpaitou Dam brings about significant improvement in the water quality of the Pihe River (a tributary of the Huai River) downstream from Lu'an City. Thus, good design and operation of large dams have to some extent mitigated water pollution in the Huai River Basin.

3.1.6 Hai River Basin

The Hai River flows through Beijing and Tianjin in northern China before emptying into the Yellow Sea at the Bohai Gulf. This river has nine large tributaries including the Zhangwei, Ziya, Daqing, Yongding, Chaobai, Beiyun, Jiyun, Tuhai, and Majia Rivers. The total length of the river, originating from the source of the longest tributary, is 1,329 km and the length of the mainstream of the Hai River in Tianjin is about 70 km (Fig. 3.7). Its annual flow is only half that of the Yellow River, or 1/30

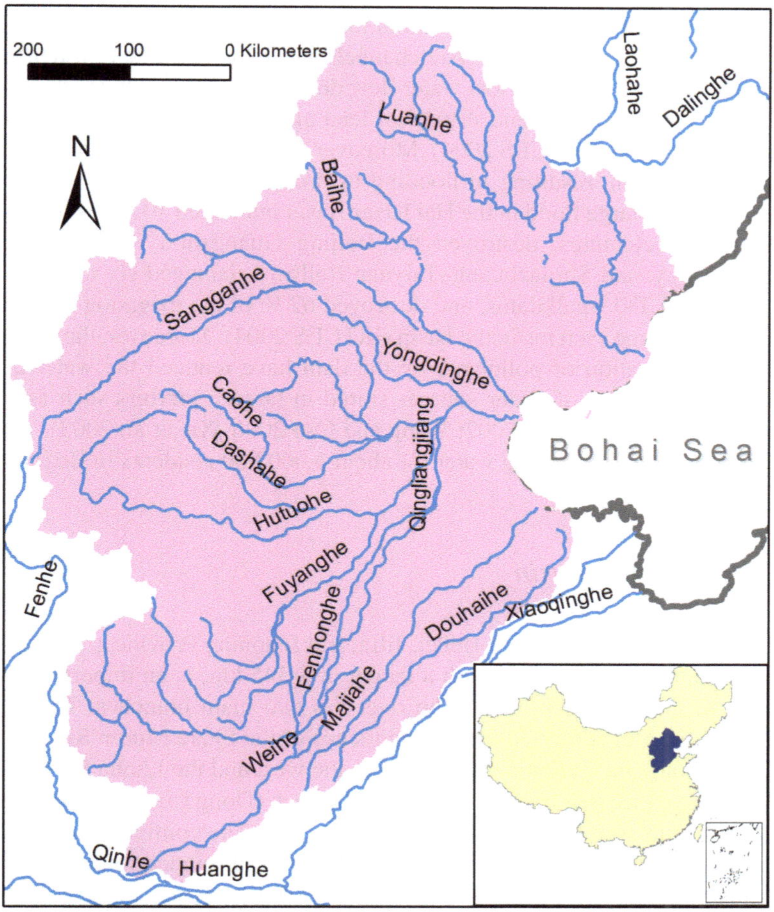

Fig. 3.7 Map of Hai River Basin

that of the Yangtze River (The Ministry of Water Resources 2013). The Hai River Basin has an area of approximately 319,000 km^2. The major tributaries of the Hai River can be divided into three reaches: the mountainous, transitional, and plains. Among them, the Yongding River is the most sediment-laden tributary, and Beijing (the capital of China) is located on this tributary. Sedimentation in the Yongding River channel in the past turned it into a perched river, making flood control for downstream cities such as Beijing very difficult. In1954, Guanting Dam was built in the upper reaches of the Yongding River for flood control and water supply. Since then, a great number of large dams have been built in the tributaries to control floods following the principle to build dams "to impound flood water in the reservoirs in the mountainous regions, to harness river channels in the middle reaches, and to discharge floods rapidly in the lower reaches" (IRTCES 2004).

However, construction of large dams in the Hai River Basin has caused serious sedimentation in the downstream river channels and estuaries. Due to water storage in the upstream reservoirs, the low water flow does not move sediments through the downstream river channels and estuaries, leading to heavy sedimentation which causes them to dry up (IRTCES 2004). Moreover, large dams have changed the flow regimes downstream, resulting in flooding or drought problems. For example, in August of 1963, a large flood in the Hai River Basin inundated 107 counties belonging to seven prefectures, destroyed the Beijing–Guangzhou Railway, Beijing–Dezhou Railway, and Shijiazhuang–Taiyuan Railway, damaged six medium-sized dams, collapsed 330 small dams, and destroyed 62 % of the irrigation systems and 90 % of the drainage systems in the basin (IRTCES 2004). Moreover, the accumulation and magnification of pollutants by the dam have reduced the water quality, threatening the quality of water sources stored in large reservoirs such as Miyun Reservoir (Wang et al. 2001, 2002; Wang and Cao 2006; Xie et al. 2003), which is the primary source for drinking water for about 3,000,000 residents in Beijing City.

3.1.7 *Liao River Basin*

The Liao River, flowing through Hebei, Jilin, and Liaoning Provinces and the Inner Mongolia Autonomous Region with a length of 1,345 km, is an important river in Northeast China (IRTCES 2004). The river has two main branches. The western branch (Xi Liao He) is formed by the confluence of the Shara Muren River flowing from Inner Mongolia Autonomous Region in the west and the Laoha River flowing from Hebei Province in the south. The eastern branch (Dong Liao He) rises from the low mountains in Jilin Province in the east (Fig. 3.8). The confluence of these two branches flows across a vast plain in central Liaoning Province to the Bohai Gulf (The Ministry of Water Resources 2013). Like the Yellow River, the Liao River has an exceedingly high sediment load because the river flows through many parts of the loess plains. Similarly, high sedimentation in the river makes flood control in downstream areas very difficult. Therefore, most dams were built to control flooding with the first dam completed in 1942 and the last one in the 1980s. At present, there are 688 dams of various sizes with a total reservoir storage capacity of 13.8 billion m^3, but the 17 are large reservoirs hold most of the water (13.2 billion m^3) (IRTCES 2004)

Dams in the Liao River Basin work better in flood control in contrast to those in the Hai River Basin. Runoff and flood peaks have been reduced by one half and the sediment load reduced by 16 % (IRTCES 2004). However, large dams have changed the flow regime and sediment loads of the river below the reservoirs (IRTCES 2004), which may impact downstream ecosystems. Moreover, the dams may block and accumulate pollutants in the reservoirs negatively affecting water quality. A GIS-based analysis showed that over 60 % of the pollution loading by nonpoint sources is concentrated in the mainstream of the Liao River (Wen et al. 2011) and the densely dammed areas in the Liao river Basin. Although it is not clear how much

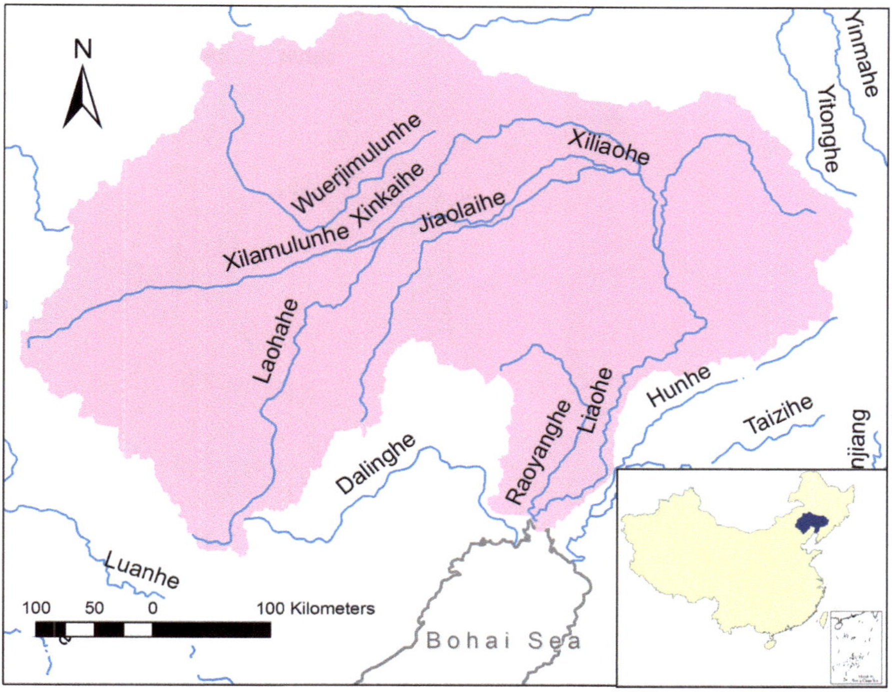

Fig. 3.8 Map of the Liao River Basin

the poor water quality is related to the large dams, serious water pollution has turned the Liao River, "Mother River" of Liaoning Province into a "sewage ditch." (Xinhuanet News Domestic 2009).

3.1.8 Songhua River Basin

The Songhua River, flowing about 1,434 km from the Changbai Mountains through Jilin and Heilongjiang Provinces, is the largest tributary of the Amur River (Heilong Jiang in Chinese) (Fig. 3.9). It joins the Amur River at the town of Dongjiang. It is another important river in Northeast China, with a basin area of about 1,443,100 km^2. The largest tributary of this river, Second Songhua River, has three dams, the Baishan, Hongshi, and Fengman Dams, which are used for the production of hydro-electricity (The Ministry of Water Resources 2013).

The impacts of large dams on the flow regime, water quality, and river ecosystems in the basin are poorly recorded. However, one study found that nonpoint source pollutants, such as nitrogen and suspended solids from agriculture and urban runoff, have increased from 1990s to 2000s, as compared to heavy metals and toxic

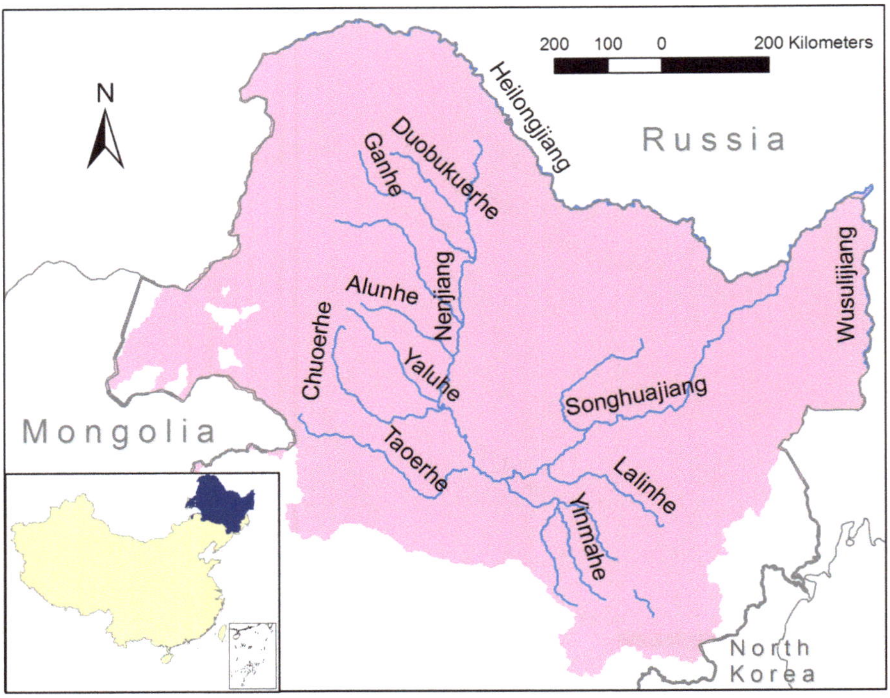

Fig. 3.9 Map of the Songhua River Basin

organic chemicals (Yu et al. 2003). This dynamic may be partially related to the effects of dams on the downstream flow regime, sedimentation, and storage capacity of the reservoir. Most importantly, the contamination of benzene in the river in November, 2005 led to a shutdown of Harbin's water supply and threats of a Russian lawsuit against China (People's com 2005).

3.2 Effects of Large Dams on Biological Diversity and Ecological Integrity

As we have discussed, it is well documented that dams can cause diverse environmental problems such as changes in the flow regime of a river, increased sedimentation in the reservoir, reduced water quality, habitat degradation, and losses in biodiversity. However, few studies have examined the impacts of the large dams on biological diversity at watershed scale and ecological integrity at the ecosystem level. In fact, biological diversity and ecological integrity are the most important indicators to reflect the environmental impacts at ecosystem or watershed levels (Li et al. 2012a, c). Below, we present frameworks for assessing the impacts of dams on the biological diversity at watershed scale and impacts on ecological integrity of rivers at ecosystem scale. In Sect. 3.3, we present case studies that assess the impacts of dams on the biological integrity and ecological integrity of the Upper-Mekong River.

3.2.1 Assessing Biological Diversity at Watershed Scale

Loss of biodiversity associated with dam construction and operation is a growing environmental concern. Although a lot of research has focused on mitigating the negative impacts of large dams on biodiversity, these studies have focused mostly on endangered, endemic, or economically valuable fish species and rarely on other aquatic and terrestrial species. In most cases, these studies have investigated the impacts of dams on biodiversity at the species level and have overlooked the impacts on biodiversity at ecosystem and watershed/landscape levels. Indeed, assessment of large-scale biodiversity losses associated with dam construction and operation is very important if upstream–downstream relationships and biological interactions are considered.

In Article 2 of the Convention on Biological Diversity, biological diversity is defined as "variability among living organisms from all sources including inter alia, terrestrial, marine and other aquatic ecosystems and the ecological complexes of which they are part; this includes diversity within species, between species and of ecosystems." Hence, biodiversity is not only the sum of all ecosystems, species, and genetic material, but also includes the variability within and among them. Biological diversity is often understood at three levels: genetic, species, and ecosystem diversity. All three levels of diversity are necessary for the continued survival of all life including humans (Primack 2012).

Previous studies (Whittaker 1972; Colwell 2009; Wilson 1988) have developed various indices to assess biological diversity. Although these indices are appropriate in disturbed or natural environments, few indicators have been created for assessing the impacts of large dams on the biological diversity at a landscape or region level. Recently, we developed the Vegetation Impact Index (VII) for assessing the impacts of large dams on plants at the community level (Li et al. 2012a). The VII was developed as a quantitative index to predict the degree to which vegetation types would be affected by rising water levels and inundation calculated as follows:

$$\textit{Riparian vegetation}: VII = R_a / R_r$$
$$\textit{Upland vegetation}: VII = R_b / R_u$$

where R_a is the riparian vegetation distribution range (the boundary at which the vegetation cannot be affected by the backwater) after dam construction, R_r is the riparian vegetation distribution range before dam construction; R_b is the upland vegetation distribution range (altitude) above normal water level in the reservoir region, and Ru is the upland vegetation distribution range (altitude) in the area far from dam. The values of VII are divided into five classes: (1) <0.20 is grade 5, representing

Table 3.1 Ecological risk categories for species derived from the changes of the summed dominance ratio (SDR)

Changes of species dominance (%)	Level of ecological risk at species level	
	Dominant species	Nondominant species
100	IV	III
>64	IV	II
34–64	II	I
<34	I	0

Note: 0:, no ecological risk/extremely low ecological risk; I, low ecological risk; II, medium ecological risk; III, high ecological risk; IV, Extremely high ecological risk

severe impact; (2) 0.20<0.40 is grade 4, representing heavy impact; (3) 0.40<0.60 is grade 3, representing moderate impact; (4) 0.60<0.80 is grade 2, representing light impact; and (5)>0.80 is grade 1, representing little or no impact.

We also developed an indicator of ecological risk (ER) of plant species using changes in the summed dominance ratio (SDR) to evaluate impacts of large dams at the species level (Li et al. 2012a). The SDR is calculated based on the relative coverage (C) and relative frequency (F) of a species as follows:

$$SDR(\%) = \frac{C' + F'}{2}, \quad C'(\%) = \frac{C_i}{C_I} \times 100, \quad F'(\%) = \frac{F_i}{F_I} \times 100$$

where the variables C' and F' are a species cover ratio and species frequency ratio. C_i and C describe species i cover and maximum species cover in the plant community, and F_i and F_I are species i frequency and maximum species frequency in the plant community, respectively. The ER of the species is categorized into four levels for both dominant and nondominant species based on changes in the SDR of that species (Table 3.1).

3.2.2 *Assessing Biological Integrity at Ecosystem Scale*

The concept of biological integrity originated in the US Federal Clean Water Act of 1972, which addressed concerns about the integrity of the nation's waters. Integrity refers to an unimpaired condition, and biological integrity is defined as how "pristine" an environment is by examining the biological composition, structure, and function relative to the potential or original state of an ecosystem before human disturbances were introduced (Angermeier and Karr 1994; Karr 1991).

The widely accepted definition of biological integrity is adapted from Frey (1977): "the capability of supporting and maintaining a balanced, integrated, adaptive community of organisms having a species composition, diversity, and functional organization comparable to that of the natural habitat of the region." The implications of this definition are that living systems have a variety of scales relative to which they exist, that one can quantify the parts that sustain or contribute to a

Table 3.2 Selected metrics for the modified fish index of biological integrity (F-IBI) in the middle reaches of the Upper-Mekong River

Categories	Selected metrics	Abbreviation	RE
Species composition and richness	Number of species	SM_1	+
	Number of native species	SM_2	+
	Proportion of native species	SM_3	+
	Proportion of endemic species	SM_4	+
Trophic guilds	Proportion omnivorous	SM_5	–
	Proportion invertivorous	SM_6	+
	Proportion herbivorous	SM_7	+
	Proportion planktivorous	SM_8	–
Habitat	Proportion of lotic habitat (includes water column and benthic habitat)	SM_9	+
Tolerance	Proportion of tolerance individuals	SM_{10}	-

Note: RE means response to environmental condition; "+" means positive correlation to pristine environmental conditions; "–" means negative correlation)

system's functioning, and that all systems must be seen within the context of their environments and evolutionary histories.

Karr (1981, 1986) created the index of biological integrity (IBI) as a means for resource managers to quantify the biological integrity of aquatic systems. After that, many different indices have been developed for various organisms and ecosystems, but none have been developed specifically to assess the impacts of dams on the biological integrity of rivers. As such, we developed a modified fish-IBI (F-IBI) for evaluating the health of the aquatic ecosystem disturbed by large dams in the middle reaches of the Upper-Mekong River. We used ten metrics that can be grouped into four categories: species composition and richness, trophic guilds, fish habitat, and tolerance to disturbance. Species composition and richness included four vectors: the number of species, number of native fish, proportion of native fish, and proportion of endemic fish. The trophic guilds of fish assemblages were classified into four vectors: omnivorous, invertivorous, herbivorous, and planktivorous. Fish habitat was classified as the proportion of the total habitat that is a lotic system; this was selected as the key vector because most fish in this section of the Upper-Mekong River live in rapid flowing waters. For tolerance, the proportion of tolerant fish was used as the key vector as tolerant species are more adaptable to varying environmental condition, e.g., insensitive to changes of flow regimes and water quality associated with dam inundation and operation (Li et al. 2013).

Changes in the F-IBI at both temporal and spatial scales reflect ecological variation from pristine to degraded aquatic ecosystems induced by dam impoundment and flow regulation (Li et al. 2013). The index value of each sampled site can be normalized on a scale of 0–100 based on the ten selected metrics (SM1–SM10).

When the response to environmental conditions (RE in Table 3.2) is positive, the SM can be normalized as follows:

$$SM = \frac{V - V_{\min}}{V_{\max} - V_{\min}} \times 100$$

Table 3.3 Categories of aquatic ecosystem health derived from modified F-IBI scores

F-IBI score	Level	Assessment	Character
81 ≤ 100	I	Excellent	Relative pristine condition, diverse native fish species
61 ≤ 80	II	Good	Decreased native fish species, low disturbance from human activities
41 ≤ 60	III	Fair	Dramatically decrease in native species, moderate disturbance from human activities
21 ≤ 40	IV	Poor	Few native fish species, fish habitats highly impacted by impoundment and flow regulation
1 ≤ 20	V	Bad	Dominance by exotic fish species, no native fish species.
0	VI	No fish	No fish caught in the waterways.

When the response to environmental conditions (RE in Table 3.2) is negative, the SM can be normalized as follows:

$$SM = 100 - \frac{V - V_{\min}}{V_{\max} - V_{\min}} \times 100$$

The score of the modified F-IBI for each site is then calculated as the average of the 10 SM values:

$$Modified\ F - IBI = \sum_{i=10}^{i=1} SM_i \,/\, 10$$

where V is the value of a sampled site, $V_{\min}$ is the minimum value of all sampled sites, $V_{\max}$ is the maximum value of all sampled sites, SM_i is the score of the sampled metric, i. The final modified F-IBI scores range from 0 to 100, which reflects different levels of aquatic ecosystem health (Table 3.3).

Similarly, we currently are developing a modified phytoplankton IBI for evaluating the health of the phytoplankton communities impacted by the large dams in middle reaches of the Upper-Mekong (Li et al. 2012b).

3.3 Case Study: Environmental Impacts of Dams on the Upper-Mekong River

3.3.1 Impacts on Biological Diversity

3.3.1.1 Vegetation Impacts

We conducted field surveys to study the impacts of the large dams on vegetation distribution patterns in riparian and upland areas along the middle-lower reaches of the Upper-Mekong River during the dry seasons from 2004 to 2010. In total, 24 transects perpendicular to the river channel were set up to sample the riparian and

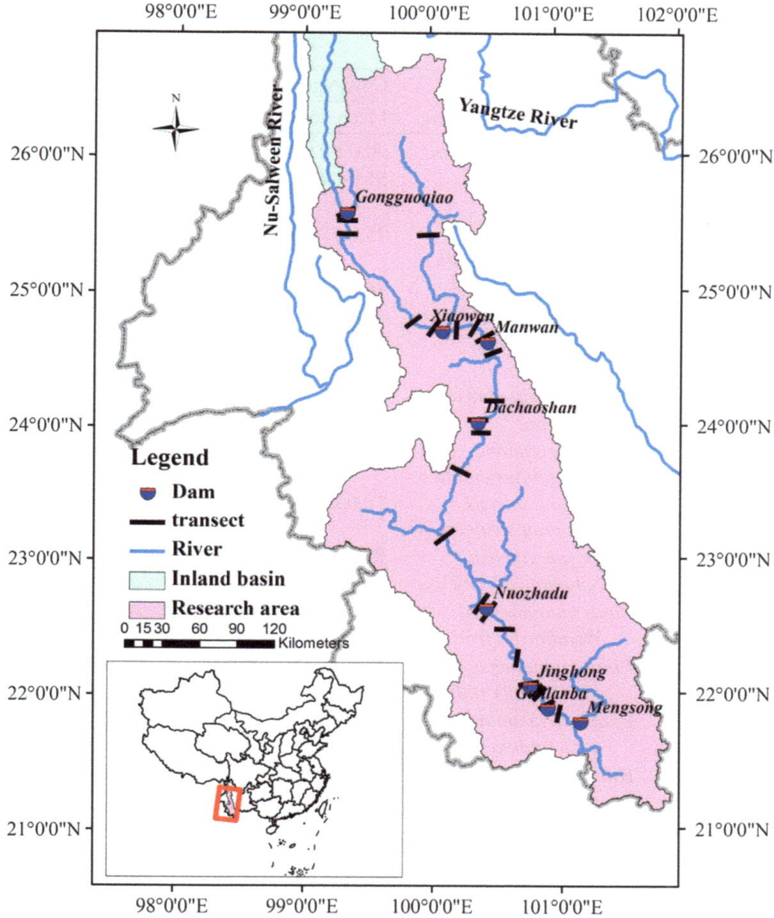

Fig. 3.10 Map of cascading dams along the middle-lower reaches of the Upper-Mekong River and location of transects used to sample vegetation

upland vegetation on the upstream impoundment banks and downstream river banks of eight cascading dams, the Gongguoqiao, Xiaowan, Manwan, Dachaoshan, Nuozhadu, Jinghong, Ganlanba, and Mengsong dams (Fig. 3.10). Along each transect, 5–6 $\times$ 10 m^2 quadrats were arrayed along the elevational gradient and a total of 126 quadrats were used to sample the vegetation in the study area. Using the reservoir water level as a reference, the number of quadrats that were inundated or will be inundated was recorded on each sampling transect, and the percentage of inundation calculated. The vegetation impact index (VII) at different sampling sites was calculated using the methods introduced in Sect. 3.2.

The vegetation was classified into 21 community types (Table 3.4). The upland vegetation was dominated by trees and shrubs whereas the riparian areas supported

Table 3.4 Vegetation types along the middle-lower reaches of the Upper-Mekong River Basin measured before dam construction

	Vegetation types	Dominant plant species	Coverage Mean ± SD	Habitat	Quadrats T	I	Per. I. (%)
TC	I	*Dendrocalamus strictus*	90.0 ± 0.0	Upland	4	1	25.00
	II	*Ficus semicordata*	55.0 ± 7.1	Upland, Riparian	2	0	0.00
	III	*Ficus semicordata*, *Phyllanthus emblica*	75.0 ± 21.8	Upland, Riparian	9	3	33.33
	IV	*Dalbergia obtusifolia*, *Wendlandia scabra*	76.6 ± 16.3	Upland	19	5	26.32
	V	*Pinus khasya* var. *langbianensis*, *Dalbergia obtusifolia*	81.0 ± 16.7	Upland	15	0	0.00
	VI	*Engelhardtia colebrookiana*, *Phyllanthus emblica*	71.1 ± 24.0	Upland	14	5	35.71
	VII	*Pinus yunnanensis*, *Castanopsis delavayi*	70.0 ± 12.6	Upland	7	0	0.00
	VIII	*Pinus yunnanensis*, *Schima wallichii*	83.3 ± 11.6	Upland	3	0	0.00
	IX	*Pinus yunnanensis*, *Woodfordia fruticosa*	71.3 ± 15.5	Upland	4	0	0.00
	X	*Pinus yunnanensis*, *Phyllanthus emblica*	67.8 ± 27.6	Upland	13	1	7.69
SC	XI	*Bauhinia acuminata* var. *candida*	80.0 ± 17.3	Upland	3	3	100.0
	XII	*Homonoia riparia*	68.8 ± 16.5	Riparian	4	2	50.00
	XIII	*Woodfordia fruticosa*, *Homonoia riparia*	50.0 ± 0.0	Riparian	1	1	100.0
	XIV	*Picrasma quassioides* var. *quassiodes*, *Broussonetia papyrifera*	91.3 ± 2.5	Upland	4	1	25.00
	XV	*Dendrocalamus strictus*, *Broussonetia papyrifera*	73.3 ± 28.9	Upland	3	2	66.67
	XVI	*Buddleja asiatica*	55.0 ± 12.9	Upland	4	3	75.00
	XVII	*Rapanea neriifolia*, *Ficus tikoua*	65.0 ± 16.6	Riparian	5	2	40.00
GC	XVIII	*Vernonia patula*, *Digitaria sanguinalis*	67.5 ± 35.2	Riparian	4	2	50.00
	XIX	*Equisetum diffusum*	25.0 ± 0.0	Riparian	1	1	100.0
	XX	*Buddleja* sp., *Eupatorium odoratum*	66.7 ± 20.8	Riparian	3	2	66.67
	XXI	*Phragmites* sp.	48.8 ± 11.8	Riparian	4	1	25.00
					126	35	27.78

TC tree community, *SC* shrub community, *GC* grass community, *T* total number of quadrats sampled for each vegetation, *I* the number of quadrats that will be inundated, *Per.I.* percentage of inundated quadrats

Table 3.5 Vegetation impact index (VII) associated with dam construction

Reservoir areas	Vegetation type	VII	Grade	Vegetation type	VII	Grade
GGQ	VII	0.0723	5	XVIII	0.0000	5
XW	XVI	0.5503	3	XVIII	0.0079	5
	IV	0.5059	3	VII	1.0000	1
	III	0.0972	5	VIII	1.0000	1
	VI	0.0769	5	X	0.5543	3
MW	IV	0.8358	1	VIII	1.0000	1
	VI	0.7295	2	X	1.0000	1
	V	0.6489	2	IX	1.0000	1
DCS	IV	0.8770	1	VI	0.8117	1
	III	0.8145	1	V	0.7680	2
NZD	III	0.6847	2	II	0.2553	4
	XV	0.0000	5	VI	0.7385	2
	XXI	0.0141	5			
JH	II	1.0000	1	XX	0.0000	5
	I	0.3057	4	XII	0.0000	5
	XIV	0.5000	3	XIII	0.0000	5
	XV	0.8072	1	III	0.9552	1
	XIX	0.0000	5	IV	1.0000	1
	XXI	0.0000	5	V	1.0000	1
GLB	XI	0.0000	5			
	XIV	1.0000	1	XXI	0.1538	5
	XII	0.1538	5	V	1.0000	1
	XVIII	0.1538	5	XX	1.0000	1
	XXI	0.0833	5	I	1.0000	1
MS	XII	0.0833	5			

GGQ GongGuoqiao Dam, *XW* Xiaowan Dam, *MW* Manwan Dam, *DCS* Dachaoshan Dam, *NZD* Nuozhadu Dam, *JH* Jinghong Dam, *GLB* Ganlanba Dam, *MS* Mengsong Dam

grasses and shrubs. Eight of the ten tree communities identified in our analysis were distributed in upland areas whereas two were found on upland and riparian areas. Of the shrub communities, three of the seven were distributed in riparian areas and the other four in upland sites. All of the grass communities were distributed in riparian areas (Table 3.4). The vegetation has the potential to be greatly impacted by the rising water levels following dam operation. When the water levels increase with dam operation, 36 out of 126 quadrats were/will be inundated by the reservoirs (Table 3.4). The grass and shrub layers were/will be more seriously affected than the tree layer (Li et al. 2012c).

The calculated vegetation impact index (VII) of each community type associated with dam construction shows that Gongguoqiao Dam severely impacted vegetation type VII, *Pinus yunnanensis*, *Castanopsis delavayi* and type XVIII, *Vernonia patula*, *Digitaria sanguinalis* (Table 3.5). The Xiaowan Dam severely affected vegetation type III, *Ficus semicordata*, *Phyllanthus emblica*, type VI, *Engelhardtia colebrookiana*, *Phyllanthus emblica*, and type XVIII, *Vernonia patula*, *Digitaria sanguinalis* and moderately impacted vegetation type XVI, *Buddleja asiatica*, and type IV, *Dalbergia obtusifolia, Wendlandia scabra*. The Manwan Dam only lightly impacted vegetation type V, *Pinus khasya* var. *langbianensis, Dalbergia obtusifolia* and type VI, *Engelhardtia colebrookiana, Phyllanthus emblica*. The Dachaoshan Dam lightly affected vegetation type V, *Pinus khasya* var. *langbianensis, Dalbergia obtusifolia*. The Nuozhadu Dam severely impacted vegetation type XV, *Dendrocalamus strictus, Broussonetia papyrifera*, and type XXI, *Phragmites* sp., heavily impacted vegetation type II, *Ficus semicordata*, and lightly impacted vegetation type III, *Ficus semicordata, Phyllanthus emblica*, and type VI, *Engelhardtia colebrookiana, Phyllanthus emblica*. The Jinghong Dam severely affected vegetation type XIX, *Equisetum diffusum*, type XXI, *Phragmites* sp., type XI, *Bauhinia acuminata* var. *candida*, type XX, *Buddleja* sp., *Eupatorium odoratum*, type XII, *Homonoia riparia*, and type XIII, *Woodfordia fruticosa, Homonoia riparia*, heavily affected vegetation type I, *Dendrocalamus strictus*, and moderately affected vegetation type XIV, *Picrasma quassioides* var. *quassiodes, Broussonetia papyrifera*. Ganlanba Dam severely affected vegetation type XII, *Homonoia riparia*, type XVIII, *Vernonia patula, Digitaria sanguinalis*, and type XXI, *Phragmites* spp. The Mensong Dam severely affected vegetation type XXI, *Phragmites* sp. and type XII, *Homonoia riparia* (Li et al. 2012c).

3.3.1.2 Ecological Risk of Plant Species

We conducted surveys of key species in the Manwan and Xiaowan reservoir areas in 1997, before construction of the Xiaowan Dam, and in 2011, after the Xiaowan Dam was constructed to quantify species dominance and their ecological risk imposed by the dams. Three plots were established along a 10 km section from the Xiaowan to Manwan dam: $15 \times 10\text{-m}^2$ quadrats, five per plot, were used to sample trees; the dominant shrub species were sampled in three 5-m^2 sub-quadrats in quadrat; and dominant herbs species were recorded three 1-m^2 sub-quadrats placed in each quadrat. The number and total cover of the dominant species in each sub-quadrat were recorded (Fig. 3.11).

The results indicated that at sampling site No. 1 that, before the construction of Xiaowan Dam in 1997, *Phyllanthus emblica* and *Woodfordia fruticosa* were the dominant tree and shrub species but were replaced by *Engelhardtia colebrookeana* after dam construction in 2010. The disappearance of the herb *Heteropogon contortus* and appearance of *Capillipedium assimile* and *Ageratina adenophora* were the major changes in the herbaceous vegetation at this site. At sampling site No. 2, the dominant tree species were *Castanopsis delavayi* and *Castanopsis fleuryi* before dam construction, but *Pinus yunnanensis* became the dominant species post dam construction and *Castanopsis fleuryi* vanished. In the shrub and herb layers, *Glochidion daltoni* and *Pogonatherum paniceum* became the dominant species after construction of the Xiaowan Dam. At sampling site No. 3, the dominant species were crops, such as *Bauhinia acuminata, Dalbergia obtusifolia,* and *Oplismenus compositus*, before the dam, and the invasive plants, *Ageratina adenophora* and *Chromola enaodoratum*, became dominant when farmland was abandoned (Table 3.6).

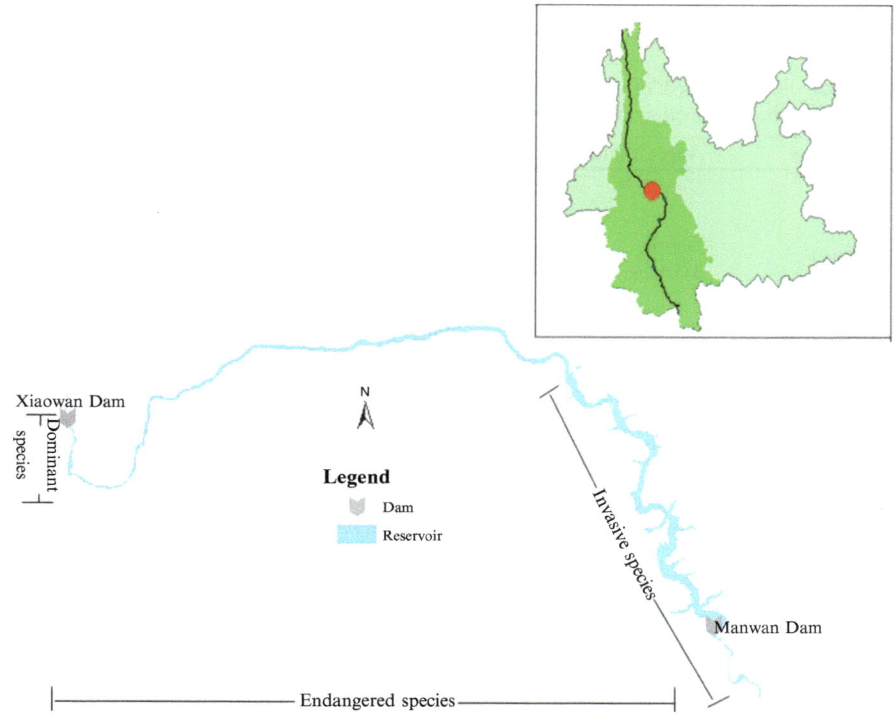

Fig. 3.11 Map of sampling sites for assessment of ecological risk of vegetation along the Xiaowan and Manwan reservoirs in the middle reaches of the Lancang River

Table 3.6 Changes in key species of riparian vegetation at three sampling sites before (1997) and after (2010) construction of the Xiaowan Dam

			SDR (mean ± SE)	
Sample site	Vegetation layer	Plant species	1997	2010
1	Tree	*Phyllanthus emblica*	100 ± 1	79 ± 4
		Engelhardtia colebrookeana	13 ± 10	100 ± 1
	Shrub	*Woodfordia fruticosa*	100 ± 1	52 ± 24
	Herbs	*Heteropogon contortus*	100 ± 10	0
		Capillipedium assimile	39 ± 9	67 ± 25
		Eupatorium adenophorum	0	75 ± 25
2	Tree	*Castanopsis delavayi*	55 ± 15	69 ± 19
		Castanopsis fleuryi	100 ± 20	0
		Pinus yunnanensis	0	100 ± 0
	Shrub	*Melastoma normale*	88 ± 9	29 ± 4
		Glochidion daltoni	85 ± 15	100 ± 24
	Herbs	*Pogonatherum paniceum*	43 ± 4	65 ± 0
3	Tree	*Dalbergia obtusifolia*	77 ± 5	0
		Bauhinia acuminate	100 ± 4	0
	Herbs	*Ageratina adenophora*	16 ± 3	79 ± 12
		Chromolaena odoratum	0	100 ± 12
		Oplismenus compositus	100 ± 25	0

Table 3.7 Ecological risk of plant species at sampling sites of No. 1, 2, and 3 before (1997) and after (2010) the construction of Xiaowan Dam

Sample site	Vegetation layer	Dominant species	Risk level
1	Tree	*Phyllanthus emblica*	I
		Engelhardtia colebrookeana	III
	Shurb	*Woodfordia fruticosa*	II
	Herbage	*Heteropogon contortus*	IV
		Capillipedium assimile	I
		Ageratina adenophora	IV
2	Tree	*Castanopsis delavayi*	I
		Castanopsis fleuryi	IV
		Pinus yunnanensis	IV
	Shurb	*Glochidion daltoni*	I
		Melastoma nrmale	III
	Herbage	*Pogonatherum paniceum*	I
3	Tree	*Dalbergia obtusifolia*	IV
		Bauhinia acuminate	IV
	Herbage	*Chromolaena odoratum*	IV
		Ageratina adenophora	II
		Oplismenus compositus	IV

Note: 0, no ecological risk/extremely low ecological risk; I, low ecological risk; II, medium ecological risk; III, high ecological risk; IV, extremely high ecological risk

It is evident from these results that dam construction can have major impacts on the vegetation and biodiversity leading to a decrease or even loss of the dominant species and an increase in nondominant and invasive plants in the riparian zones (Li et al. 2012a).

The ecological risk analysis presented in Table 3.7 indicates that at sampling site No. 1, the tree species, *Engelhardtia colebrookeana,* was at high ecological risk (III), the shrub species, *Woodfordia fruticosa* and *Eupatorium adenophorum*, were at medium (level II) and extremely high risk levels (level IV), respectively, and the herbaceous species, *Phyllanthus emblica* and *Capillipedium assimile*, were at low ecological risk (level I). At sampling site No. 2, the tree species *Castanopsis fleuryi* and *Pinus yunnanensis* were at extremely high ecological risk (level IV), and the herb, *Pogonatherum paniceum*, was at low ecological risk (level I). At sampling site No. 3, the tree species *Dalbergia obtusifolia*, *Bauhinia acuminata,* and *Oplismenus compositus* were at the extremely high risk (level IV), the herb species, *Ageratina adenophora* and *Chromolaena odoratum*, were at moderate (level II) and extremely high levels of ecological risk (level IV). These results indicate that each plant species responded differently to dam construction and were at different levels of risk. The dominant species in some of the plant communities were at risk of being completely lost from the system whereas exotic species were poised to become dominant in some sites (Li et al. 2012a).

3.3.2 *Impacts on Ecological Integrity*

3.3.2.1 Impacts on the Biological Integrity of Fish Assemblages

We monitored changes in the fish assemblages at the Xiaowan Dam at both temporal and spatial scales. The surveys were conducted in 2008, 2010, and 2011 that corresponded to the periods of before dam impoundment, during water storage and after dam operation. Because the Xiaowan Dam is far upstream from the Manwan Dam and there were no dams further upstream from the Xiaowan Dam, the 2008 fish survey was regarded as the baseline sample that represented pristine conditions. The 2010 survey assessed the impacts of water storage (as the reservoir began filling) on fish assemblages, and the 2011 survey was the impact of the fully operational dam on the fish communities. We selected four sampling sites to monitor the fish populations (Fig. 3.12). Sample site S1 was located below the reservoir and could assess the impacts of flow regulation on the fish communities. Site S2 was located at the head of the reservoir in static water with a maximum depth 252 m after dam impoundment; S3 was located in static water in the middle reach of the mainstream reservoir region; and S4 was located in a tributary reservoir region of the lower Heihuijiang River, representing the effects of dam impoundment and operation on fish populations in tributaries.

The sampling sites were selected to represent a range of habitat and hydrological conditions that would be differentially impacted by the dam and its operation. The fish index of biological integrity (F-IBI) was calculated at each site at each time period to assess changes in the health of the river ecosystem before and after the dam became fully operational.

The calculated F-IBI for the different sites shows that before dam impoundment in 2008, the F-IBI score decreased from 99.83 (level I) in the upstream site, S3, to 70.59 downstream at site S2 (level II). The F-IBI score at S4 in the tributary of was 48.86 (level III), much lower than that of the mainstream. With the impoundment of the Xiaowan Dam in 2010, the scores of F-IBI decreased sharply to level V at all sites. When the dam was operational and generating power in 2011, the F-IBI at S2, S3, and S4 all decreased slightly whereas S1 increased (Table 3.8). These results clearly show that the construction and operation of large dams can have negative impacts on fish assemblages. The major causes of decreased biological integrity of the river were due to loss of lotic habitat, habitat fragmentation, migration blockage, loss of native fish species, and an increase in nonnative fish species (Li et al. 2013).

3.3.2.2 Impact on the Biological Integrity of Plankton

We conducted surveys of the phytoplankton assemblages in 1988, 1997, and 2011 at nine sites (Fig. 3.13). In 1988 and 1997, the phytoplankton assemblages were sampled in the dry season only (April) whereas they were sampled twice in 2011 during both the dry and rainy season (October). The 1984 survey represents the natural pristine state of the river unaffected by dams. The 1997 survey is after the

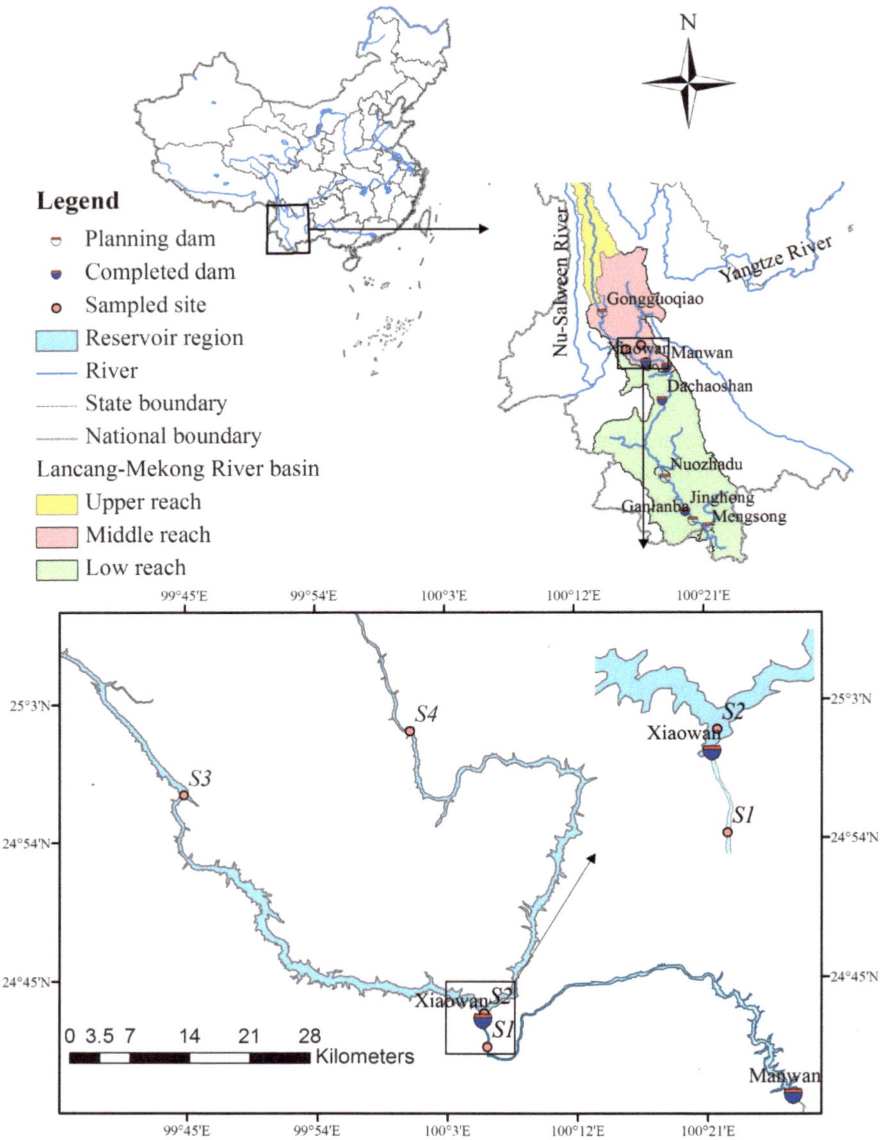

Fig. 3.12 Sampling sites for monitoring fish assemblages in the Xiaowan reservoir area

Table 3.8 Modified F-IBI scores in the Xiaowan reservoir and middle reaches of the Upper-Mekong River before (2008) and after (2010, 2011) dam construction

F-IBI	S1	S2	S3	S4	S1–S2/km	S2–S3/km	S2–S4/km
2008	NA	70.59 (II)	99.83 (I)	48.86 (III)	NA	0.53	0.33
2010	17.59 (V)	19.77 (V)	19.84 (V)	16.74 (V)	0.56	0.00	0.05
2011	26.14 (IV)	14.21 (V)	15.50 (V)	12.36 (V)	3.09	0.02	0.05

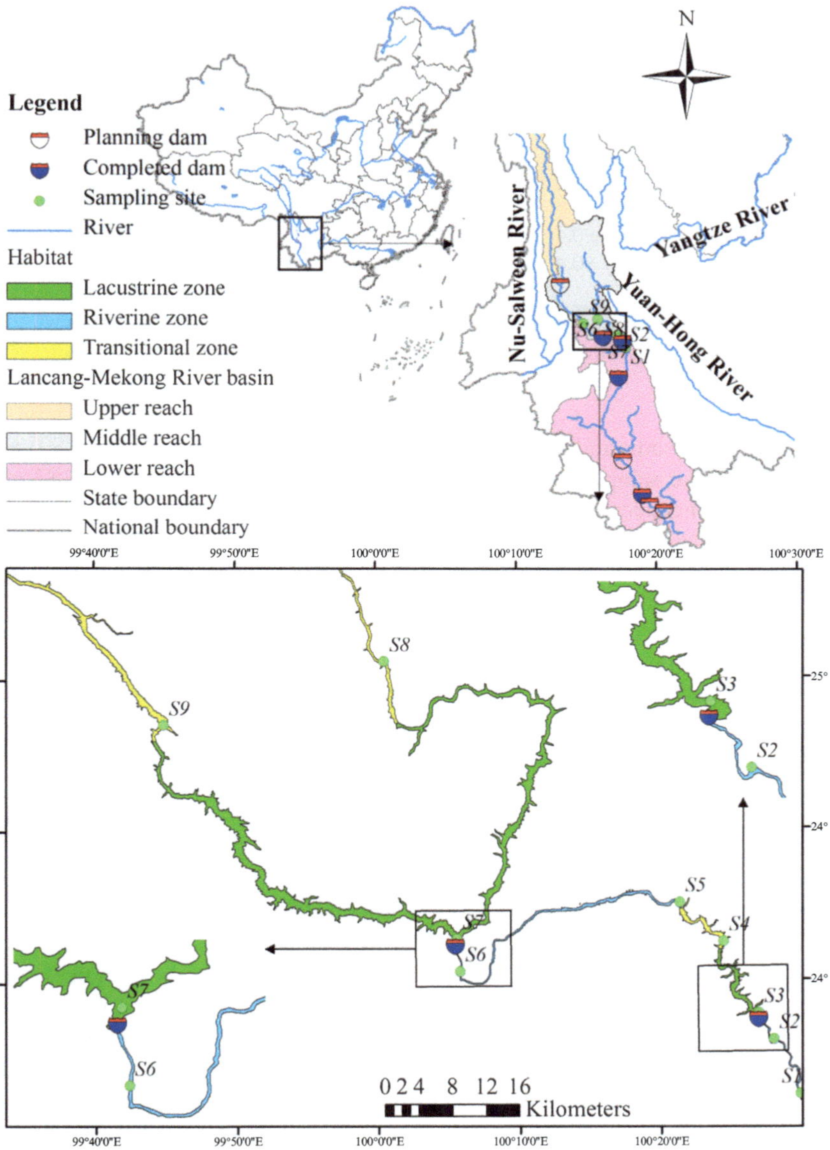

Fig. 3.13 Sampling sites for plankton assemblages in the reservoirs of three cascading dams, the Xiaowan, Manwan, and Dachaoshan dams

Manwan Dam was completed and became operational in 1995. Thus some but not all of the sampling sites were affected by dam impoundment and flow regulation in 1997. In 2011, both the upstream dam of Xiaowan (2010) and downstream dam of Dachaoshan (2003) were completed and operational, and all sampling sites were impacted by the dams. In 1984, seven sites were sampled (S1, S3, S4, S5, S7, S8,

Table 3.9 The P-IBI in the middle reaches of the Upper-Mekong River measured at four time periods: before the dams were built (April 1984), after the first dam became operational (April 1997), and after all three dams were fully operational during the dry season (April 2011) and rainy season (Oct 2011)

Sampling years	DM			MR				DX	XR			
	S1	S2	Mean±SD	S3	S4	S5	Mean±SD	S6	S7	S8	S9	Mean±SD
2011.4	2.6	3.0	2.8±0.3	2.6	2.8	2.4	2.6±0.2	3.6	2.4	3.2	2.8	2.8±0.4
2011.10	3.4	2.8	3.1±0.4	2.6	2.6	2.4	2.5±0.1	3.0	2.8	2.4	2.0	2.4±0.4
1997	2.4	3.0	2.7±0.4	2.2	2.2	3.6	2.7±0.8	NA	4.0	3.8	4.0	3.9±0.1
1988	4.6	NA	4.6	3.6	4.8	4.0	4.1±0.6	NA	4.4	4.0	3.0	4.1±0.3

DM downstream of Manwan hydropower dam; *MR* Manwan reservoir region; *DX* downstream of Xiaowan hydropower dam, *XR* Xiaowan reservoir region; *NA* not available

and S9), eight sampling sites in 1997 (S1, S2, S3, S4, S5, S7, S8, and S9) and all nine sites were sampled in 2011. After the dams became operational, S3 and S7 represented the lacustrine zone of each reservoir region; S4, S8, and S9 represented the transitional zone; S5 and S6 represented the riverine zone of the Manwan reservoir region; S1 and S2 represented the riverine zone of the Dachaoshan Dam and the downstream region below the Manwan Dam (Fig. 3.13). The planktonic index of biological integrity (P-IBI) was structured and calculated following the same procedures used for the F-IBI.

The results of P-IBI show that in 1988, before the dams had been constructed, the P-IBI in the mainstream ranged from 4.0 to 4.8 with the exception of site S3 which was lower which might have been associated with the early phases of construction of the Manwan Dam which began in 1986 (Table 3.9). After the Manwan Dam became operational in 1995, the first large dam on the mainstream of Langcang River, the P-IBI decreased both downstream and in the reservoir the Manwan Dam. In contrast, the P-IBI upstream of the Manwan reservoir increased slightly to 3.9. After all three cascading dams became operational in 2011, the P-IBI decreased sharply in the Xiaowan reservoir area during both the rainy (2.4) and dry (2.8) seasons. The P-IBIs downstream of each dam were higher than those in the impoundment areas; however, all values decreased over time as compared to the pre-dam measurements (Table 3.9). These results indicate that the health of aquatic ecosystems both upstream and downstream from large dams will be negatively affected by the construction and operation of dams due to changes in the flow regime, the loss of lotic habitat, and nutrient accumulation in the reservoirs (Li et al. 2012b).

References

Angermeier PL, Karr JR (1994) Biological integrity versus biological diversity as policy directives. BioScience 44:690–697

AsiaNews (2005) Pearl River pollution a serious concern. http://www.asianews.it/news-en/Pearl-River-pollution-a-serious-concern-3264.html. Dec 5, 2005

Baran E, Myschowoda C (2009) Dams and fisheries in the Mekong Basin. Aquat Ecosyst Health Manage 12:227–234

Barlow C, Baran E, Halls AS, Kshatriya M (2008) How much of the Mekong fish catch is at risk from mainstream dam. Catch Culture 14:16–21

BBC (2001) China builds second biggest dam http://news.bbc.co.uk/2/hi/asia-pacific/1416108.stm. July 1, 2001

Chen H, Yuan X, Chen Z, Wu Y, Liu X, Zhu D, Wu N, Zhu Q, Peng C, Li W (2011) Methane emissions from the surface of the Three Gorges Reservoir. J Geophys Res 116, D21306

China Daily (2011) Pearl River fishing ban may reduce net loss. http://www.china.org.cn/2011-04/14/content_22356145.htm. April 14, 2011

Colwell RK (2009) Biodiversity: concepts, patterns, and measurement. In: Levin SA (ed) The Princeton guide to ecology. Princeton University, Princeton, pp 257–63

Dong Q (2005) Three Gorges Dam: natural flow and dam operations. Aquat Sci 106:110–127

Dugan PJ, Barlow C, Agostinho AA, Baran E, Cada GF, Chen D, Cowx IG, Ferguson JW, Jutagate T, Mallen-Cooper M (2010) Fish migration, dams, and loss of ecosystem services in the Mekong basin. Ambio 39:344–348

Frey DG (1977) Biological integrity of water—an historical approach. The integrity of water. Proceedings of a symposium. US Environmental Protection Agency, Washington, DC, pp 127–140

Fu K, He D, Lu X (2008) Sedimentation in the Manwan reservoir in the Upper Mekong and its downstream impacts. Quaternary International 186:91–99

Grumbine RE, Xu J (2011) Mekong hydropower development. Science 332:178–179

Hays J (2009) Yellow River. http://factsanddetails.com/china.php?itemid=448

He D, Liu J, Hu J, Fen Y, Gan S (2007) Transboundary eco-security and its regulation system in the longitudinal range—Gorge region. Chin Sci Bull 52:1–9

He G, Lu Y, Mol AP, Beckers T (2012) Changes and challenges: China's environmental management in transition. Environ Dev 3:25–28

Huang Z (2001) Biodiversity conservation for the Three Gorges Project. Biodiv Sci (In Chinese) 9:472–481

IRTCES (2004) International Research and Training Center on Erosion and Sediment, Sediment issues of the Haihe and Liaohe Rivers in China. Working paper. Dec 2004 (Contract No. 861.368.4)

IUCN (2002) IUCN Red List of threatened species. www.redlist.org

Jiang Y, Xia J, Wang G, Zhao C (2009) Simulation of environmental change in response to operation of dams in Huaihe Basin. Water Sci Eng 2:27–36

Kang B, He D, Perrett L, Wang H, Hu W, Deng W, Wu Y (2009) Fish and fisheries in the Upper Mekong: current assessment of the fish community, threats and conservation. Rev Fish Biol Fish 19:465–480

Karr JR (1981) Assessment of biotic integrity using fish communities. Fisheries 6:21–27

Karr JR (1991) Biological integrity: a long-neglected aspect of water resource management. Ecol Appl 1:66–84

Kummu M, Sarkkula J (2008) Impact of the Mekong River flow alteration on the Tonle Sap flood pulse. Ambio 37:185–192

Li Y, Li R, Deng J (2000) A study of sediment transport and flood control in the middle reach of Yangtze River. J Sediment Res 3:12–19

Li S, He D, Fu K (2006) The correlations of multi-timescale characteristics of water level processes in Lancang-Mekong River. Chin Sci Bull 51:50–58

Li J, Dong S, Peng M, Li X, Liu S (2012a) Vegetation distribution pattern in the dam areas along middle-low reach of Lancang-Mekong River in Yunnan Province, China. Front Earth Sci 6:283–290

Li J, Dong S, Peng M, Yang Z, Liu S, Zhao C (2012b) Effects of cascading hydropower dams on composition, biomass and biological integrity of phytoplankton assemblages in the middle Lancang-Mekong River. (Draft)

Li J, Dong S, Yang Z, Peng M, Liu S, Li X (2012c) Effects of cascade hydropower dams on the structure and distribution of riparian and upland vegetation along the middle-lower Lancang-Mekong River. For Ecol Manage 284:251–259

Li X, Dong S, Liu S, Peng M, Li J, Zhao Q, Zhang Z (2012d) Ecological risk assessment of dam construction for terrestrial plant species in middle reach of Lancangjiang River, Southwest China. J Appl Ecol 23:2242–2248

Li J, Dong S, Peng M, Yang Z, Liu S, Li X, Zhao C (2013) Effects of damming on the biological integrity of fish assemblages in the middle Lancang-Mekong River basin. Ecol Indicators 34:94–102

Liu C, Xia J (2004) Water problems and hydrological research in the Yellow River and the Huai and Hai River basins of China. Hydrol Process 18:2197–2210

Liu S, Lu P, Liu D, Jin P (2007) Pinpointing source of Mekong and measuring its length through analysis of satellite imagery and field investigations. Geosp Inf Sci 10:51–56

Liu C, He J, Wang J (2012) Changes in sediments load of the Lancang-Mekong River and its reposes to the hydro-power development. 4th International Conference on Estuaries and Coasts, 8–11 Oct, Water Resources University, Hanoi

Lu X, Yang X, Li S (2011) Dam not sole cause of Chinese drought. Nature 475:174

Mount Holyoke College (2007) China's Three Gorge Dam. https://www.mtholyoke.edu/~vanti20m/classweb/website/environmentalimpact.html

MRC (2010) Mekong River Commission, State of basin report 2010. Vientiane, Lao PDR

New York Times (2011) China admits problems with Three Gorges Dam. http://www.nytimes.com/2011/05/20/world/asia/20gorges.html?_r=0. Released on May 19, 2011, written by Michel Wines

Nilsson C, Reidy CA, Dynesius M, Revenga C (2005) Fragmentation and flow regulation of the world's large river systems. Science 308:405–408

Ouyang W (2011) Interaction between landscape and pollution along a dam cascade. PhD Thesis, University of Twente, Enschede

Ouyang W, Hao F, Song K, Zhang X (2011) Cascade dam-induced hydrological disturbance and environmental impact in the upper stream of the Yellow River. Water Res Manage 25:913–927

People's Com (2005) Special issue on water pollution in Harbin. http://society.people.com.cn/GB/8217/55730/. Accessed June 20, 2013

People's Daily (2006). Lessons of Huaihe River water pollution. http://english.peopledaily.com.cn/200609/20/eng20060920_304506.html. By Chen Xuefei, Sept 20, 2006

People's Daily (2007) Full views of Three Gorges Project, Nov 30. http://www.mwr.gov.cn/english/20071130/88209.asp

Poulsen AF, Ouch P, Sintavong V, Ubolratana S, Nguyen TT (2002) Fish migrations of the Lower Mekong River Basin: implications for development, planning and environmental management. MRC technical paper no. 8. MRC, Phnom Penh

Primack RB (2012) A primer of conservation biology. Sinauer, Sunderland

Probe International (2012) Feverish Chinese dam building could trigger tsunami. http://journal.probeinternational.org/2012/04/03/press-release-feverish-chinese-dam-building-could-trigger-tsunami/

Qian BQ (2005) Recent evolutional processes of the Pearl River estuary and measures for utilization. Pearl River (in Chinese) 2:45–49

Ran L, Lu X, Yang X (2012) Mindset change needed for China's water crisis. Ambio 41:216–218

Revenga C (2003) Status and trends of biodiversity of inland water ecosystems. Secretariat of the Convention on Biological Diversity, Montreal

Shapiro J (2001) Mao's war against nature: politics and the environment in revolutionary China. Cambridge University Press, Cambridge

Sichuan Daily (2005) Pollution threatens the Yangtze. http://china.org.cn/english/2004/Nov/113316.htm. Translated by Li Jingrong for China.org.cn. Nov 27, 2004

Stone R, Jia H (2006) Going against the flow. Science 313:1034–1037

The Guardian (2007) 30% of Yellow river fish species extinct. http://www.guardian.co.uk/news/2007/jan/18/china.pollution, written by Jonathan Watts. Jan 18, 2007

The Ministry of Water Resources (2013) The Ministry of Water Resources of China Website. http://www.mwr.gov.cn/. Accessed June 20, 2013

Wang X, Cao L (2006) Economic approach to the control of agricultural non-pint source pollution in China—a case study of the Miyun Reservoir in Beijing. J Ecol Rural Environ (in Chinese) 22:88–91

Wang X, Li T, Xu Q, He W (2001) Study of the distribution of non-point source pollution in the watershed of the Miyun Reservoir, Beijing, China. Water Sci Technol 44:35–40

Wang X, Wang Y, Li T, He W, Hu Q, Zhang H (2002) Characteristics of non-point source pollution in the watershed of Miyun Reservoir, Beijing, China. Chin J Geochem 21:89–96

Wang G, Wu B, Wang Z-Y (2005) Sedimentation problems and management strategies of Sanmenxia Reservoir, Yellow River, China. Water resources research 41, W09417

Wang H, Yang Z, Saito Y, Liu JP, Sun X (2006) Interannual and seasonal variation of the Huanghe (Yellow River) water discharge over the past 50 years: connections to impacts from ENSO events and dams. Global Planet Change 50:212–225

Wang H, Yang Z, Saito Y, Liu JP, Sun X, Wang Y (2007b) Stepwise decreases of the Huanghe (Yellow River) sediment load (1950–2005): impacts of climate change and human activities. Global Planet Change 57:331–354

Wen Q, Chen X, Shi Y, Ma J, Zhao Q (2011) Analysis on composition and pattern of agricultural nonpoint source pollution in Liaohe River Basin, China. Proc Environ Sci 8:26–33

Weng Q (2007) A historical perspective of river basin management in the Pearl River Delta of China. J Environ Manage 85:1048–1062

Whittaker RH (1972) Evolution and measurement of species diversity. Taxon 21:213–251

Wilson EO (1988) Biodiversity. National Academy of Sciences, Washington, DC

WRI, IUCN, IWMI, RC BUREAU (2003) Watersheds of the World_CD. WRI and IUCN-The World Conservation Union, Washington DC, Gland, Switzerland and Cambridge, UK

Wu J, Huang J, Han X, Gao X, He F, Jiang M, Jiang Z, Primack RB, Shen Z (2004) The three Gorges dam: an ecological perspective. Front Ecol Environ 2:241–248

WWF (2005) Rivers at risk: dams and the future of freshwater ecosystems. http://www.panda.org/downloads/freshwater/riversatriskfullreport.pdf

Xie P, Wu J, Huang J, Han X (2003) Three-Gorges Dam: risk to ancient fish. Science 302: 1149–1151

Xie ZQ, Wu JQ, Xiong GM (2006) Conservation ecology of rare and endangered plants in the Three Gorges Reservoir Area. China Water Power Press, Beijing (in Chinese)

Xinhua News Agency (2007) Three Gorges offers clean energy, no major geological threats. http://www.mwr.gov.cn/english/20071128/88168.asp, Nov. 28, 2007

Xinhua News Agency (2011) Floods cause direct economic loss of 43.2 billion yuan in China. Retrieved 30 June, 2011, from http://news.xinhuanet.com/english2010/china/2011-06/30/c_13959118.htm

Xinhuanet News Domestic (2009) “Liaohe pollution” to achieve substantive results. http://www.sourcejuice.com/1269884/2009/10/28/Liaohe-pollution-achieve-substantive-results/ Released on Oct 28, 2009

Yu S, Shang J, Zhao J, Guo H (2003) Factor analysis and dynamics of water quality of the Songhua River, northeast China. Water Air Soil Pollut 144:159–169

Zhao Q, Liu S, Deng L, Dong S, Yang J, Wang C (2012a) The effects of dam construction and precipitation variability on hydrologic alteration in the Lancang River Basin of southwest China. Stoch Environ Res Risk Assess 26:993–1011

Zhao Q, Liu S, Deng L, Yang Z, Dong S, Wang C, Zhang Z (2012c) Spatio-temporal variation of heavy metals in fresh water after dam construction: a case study of the Manwan Reservoir, Lancang River. Environ Monit Assess 184:4253–4266

Zhong Y, Power G (1996) Environmental impacts of hydroelectric projects on fish resources in China. Regul Rivers: Res Manage 12:81–98

Chapter 4
Socioeconomic Impacts of Dams in China: Focusing on Relocated People

4.1 An Overview of Dam Relocated People in China

As the second largest economy in the world and the fastest growing major economy, China has built more large dams than any other country in order to increase irrigation, flood control, and hydropower (Holz 2008; WCD 2000). The construction of dams and reservoirs often, if not always, involves displacement of communities and livelihoods. The total number of relocated people as a result of dam construction in China is reported to be 10.2 million according to official statistics, but this number is very controversial and is much lower than some independent estimates (Wang et al. 2013a, b). According to the study of Wang and colleagues (2007a), by 2006 there were 22.8 million people relocated because of dam construction in China. This study also divided the history of dam-caused relocations into three periods, which are summarized below.

First period (1950–1957): There were about 300,000 dam-caused relocations during this period. Because it was right after the Chinese Revolution and the government controlled large amounts of land, relocation programs were relatively easy to conduct.

Second period (1958–1977): The number of relocated people increased rapidly to 4.91 million. However, due to social instability, particularly the Cultural Revolution, relocation policies were often neglected and relocated people suffered for a long time from low compensation and unreasonable living conditions.

Third period (1978–present): Most large dams in China were built during this period, which caused the most relocated people in China's history—22.8 million by 2006. Besides the need to relocate such a dramatically large population, rapidly changing socioeconomic conditions added more complexity to the relocation process.

According to the study of Tang (2009), there are four major types of dam-caused relocations in China.

P. Wang et al., *The Large Dam Dilemma: An Exploration of the Impacts of Hydro Projects on People and the Environment in China*, DOI 10.1007/978-94-007-7630-2_4,

Type 1: Agricultural-oriented relocation. This type of relocation is suitable for less developed regions, where there are few industries and population densities are relatively low. The government generally clears new farmland and provides new buildings and infrastructure for the relocated people, who usually continue to support their livelihoods through agricultural activities as they did before relocation.

Type 2: Urban and suburban area relocation. This type of relocation is suitable for industrialized regions, where there are more job opportunities, particularly in manufacture and service industries. Relocated people are not given farmland, but can work in factories or find other jobs in urban areas.

Type 3: Whole community relocation. In this type of relocation, a community is moved as a whole unit, and their political structures are maintained in the new location. The communities are relocated to either agricultural areas or urban areas.

Type 4: This type involves different combinations of types 1 and 2 that depend on different household situations. The relocation policy could be very flexible in this type of relocation, and affected people are either relocated to a far place, or just move a short distance.

The compensation policy for relocated people has also evolved along with the development of China's society. Before the 1980s, China was under an extreme central planning system, and dam-affected people were required to sacrifice their own interests for the benefit of the country; therefore, they received very low compensations for being forced to move. Then after the reform and opening policy in 1978, the government learned from mistakes in the previous period and adjusted its resettlement and compensation practices. Since local minorities were no longer expected to sacrifice all of their assets to the national interests, compensation was expanded. Further, the Central Government enacted legislation in 1991 to regulate compensation policies in large hydraulic project affected areas. But at that time, the compensation standards were still very low. In 2006, the Central Government updated this legislation and substantially raised compensation levels by more than threefold, and also began to emphasize the importance of maintaining sustainable livelihoods for relocated people (Wang et al. 2013a, b).

While previous studies contributed significantly to understanding the social impacts of large dams in China, most of them were restricted to relatively small temporal or spatial extents. In order to understand patterns and trends, we seek to analyze relocated people and compensation policy for resettlement over several decades at the national level. In this chapter, we provide a social impact analysis framework that has straightforward implications for improving the development of compensation policies in China. First, we classify the wealth of affected people into three dimensions and discuss the loss and compensation in each dimension. Then we use this framework to analyze the evolution of compensation policies for dam relocated people in China, as well as the institutional changes during the evolution. At last we discuss new coordination mechanisms to protect the rights of relocated people.

4.2 Three Dimensions of Wealth: Material, Embodied, and Relational Wealth

Social impact analysis should take into consideration different dimensions of impacts (Freudenburg 1986; Vanclay 2003), and several papers have been published using such analyses to assess the impacts of large dams (Sadler et al. 2000; Tilt et al. 2009). To comprehensively analyze both the negative and positive social impacts of large dams at different scales, and to provide recommendations for improving compensation policies, we introduce a framework of wealth analysis, and then apply it to our analyses in the next three sections (Wang et al. 2013a, b). Wealth is a multidimensional attribute (Coleman 1994). We identified three classes of wealth based on the work of Mulder and colleagues (2009):

1. Material wealth: Includes farmland, houses, livestock, cash crops, and forests. This class of wealth can be measured in monetary units.
2. Embodied wealth: Wealth that is implicitly carried by a person. In this book we focus primarily on various skills that could be used to make a living (e.g., agricultural skills, fishery and ferry skills, and business skills.)
3. Relational wealth: Includes two components, social infrastructure (a person's social network, language, customs, and traditional festivals), and physical infrastructure (transportation conditions, healthcare, and education resources). Wealth in this class is provided by the environment and benefits individuals, but is not possessed by them.

4.2.1 Losses in Three Types of Wealth

Material wealth loss is easy to measure monetarily and, in many cases, is considered the only loss experienced by relocated people. In large dam construction projects, farmland and houses inundated, or expropriated for construction sites, are the most significant losses in material wealth for most farmers.

Embodied wealth, such as agricultural skills, once mastered, is unlikely to be forgotten. However, the environment determines the effectiveness of embodied wealth. If a farmer has grown rice for decades, but is relocated to another place and given land that is only suitable for growing wheat, he will suffer a loss in his rice-growing skills. In a worse case, a farmer who loses all his land and has to find a job in a city loses all his embodied wealth in agricultural skills.

Relational wealth is also difficult to measure, but undoubtedly has significant influence on the well-being of people. A person may have developed an economic network for selling his products and buying input materials after living in a place for many years. People have relatives and friends who socially support one another, as well as their cultural identity, featured by languages, traditional festivals, and customs. Relocation can have impacts on these economic/social/cultural networks to a great extent. Loss in relational wealth might also increase difficulty of access to education and healthcare resources, or worsen transportation conditions or water and electricity supply (i.e., physical infrastructure) (Wang et al. 2013a).

4.2.2 *Compensations, Subsidies, and Opportunities in Three Types of Wealth*

Many relocated people may receive a "fair compensation" only in terms of material wealth. They might be given new houses in better conditions than their previous ones, and if the new farmland is smaller than their former farmland, they may receive monetary subsidies.

Generally, there is no explicit compensation for embodied wealth. However, the new environment could provide relocated people opportunities to gain more of this type of wealth, sometimes far exceeding their previous level. For example, the construction of a dam convenes thousands of workers, who increase the demands for goods and services and form a prosperous local market. In this new setting, many local people might quit agricultural activities and open restaurants or shops. Even though they have abandoned agricultural production and possibly may lose such skills, they gain commercial skills that could lead to higher incomes and improving their standards of living. This could be a positive impact of dam construction on members of local communities.

Relational wealth is hard to quantitatively analyze, and decision makers generally totally ignore it when designing compensation programs. Nevertheless, relocation could also bring opportunities to generate new network wealth. Construction of infrastructure may make people better able to build a larger social network and to have better access to education and healthcare resources. Such potential gains in relational wealth would likely vary greatly among individuals (Wang et al. 2013a).

4.3 The Evolution of Chinese Compensation Policy for Dam Relocated People

Due to rapid growth in energy demand and an attempt to reduce the percentage of fossil fuels in energy supply structure, China has emphasized hydropower in long-term national planning. The expectation is for expansion in the number of hydro projects in the next decade, and the associated large-scale resettlement of local populations represents a key challenge. When resettlement is pursued, compensation is recognized as providing critical support to displaced people and to the maintenance of the legitimacy of sociopolitical structures and the modernization project (Cernea 1997).

We argue that compensation policies in China have distinguishing characteristics in different time periods, and that there is a profound relationship between the evolution of compensation policies and the great social and institutional transformations occurring in China over the past 60 years. Understanding this relationship and the fundamental drivers behind the transitions informs expectations surrounding infrastructure project planning and treatment of displaced people in China's future development. More broadly, the historical analysis offers insights into China's ongoing institutional changes at a time when the achievements of economic reforms of the past 30 years are threatened by the rigidness of the political system (Wang et al. 2013b).

We identify four epochs in the period from 1949, when the People's Republic of China was founded, to the present: 1949–1977, 1978–1993, 1994–2000, and 2001 to the present. The first two breaks derive from standard definitions of Chinese political history; 1978 was the beginning of the "reform and opening-up policy," and 1993 was the official date of establishment of market economy principles in China. The third transition date, 2000, is not as clear-cut as the first two dates, perhaps because we do not yet have sufficient historical distance. We identify the year 2000 as a placeholder within a gradual process of emergence and embedding of civil society organizations and new norms of legitimacy regarding the relations between Chinese citizens and national authorities. In this sense the year 2000 is a general reference to entry into the present epoch (Wang et al. 2013b).

As did Mazmanian and Kraft (2009), we emphasize the classic institutional orders of state, market, and civil society in defining the epochs and the transitions between epochs. For us, these institutions are coordination mechanisms that support resource allocation, administration, and knowledge production, thereby advancing security and socioecological reproduction (Hollingsworth 2000). Each historical period is associated, nominally, with expansion of the prominence of one of these modes of coordination. Expanded importance of a given coordination mechanism in determining compensation policy for displacement caused by dams is the basis of our assertion of an epochal transition. In making these claims, we acknowledge that there is not a one-to-one mapping between the periods and the coordination mechanisms, and the relative significance of the coordination mechanisms is quite uneven. While there is clearly important variance across territories and social problem domains within an epoch, our contention is that there is value in attending to general institutional tendencies (Wang et al. 2013b).

We characterize the three epochal transitions in compensation policy as follows (Wang et al. 2013b):

1. "The reform and opening-up policy" in 1978 was a state-led response to instabilities and contradictions in the previous period. The government recognized the legitimacy of private claims regarding property. Individuals were no longer forced to sacrifice their well-being without compensation.
2. The market-oriented reform in 1994 expanded compensation, particularly for losses of material wealth. Compensation for displacement became an outcome of negotiations between citizens, government, and state-owned companies engaged in dam construction. The involvement of quasi-private enterprises strengthened the ability of displaced people to make compensation claims.

3. The rise of civil society and the Internet after 2000 expanded transparency, participation, and accountability in resettlement process. Compensation came to encompass not only material wealth, but also embodied and relational wealth.

4.3.1 1949–1977: People and Communities in Service to the State

During the period 1949–1977, China was organized through a central planning system. Factories, farmlands, and most of the businesses were state-owned or collective-owned. The government managed and allocated most important resources, made plans for every economic sector, and was eager to organize "great production movements." This period was characterized by construction of large-scale infrastructure across the country, including large dams. Different levels of political districts actively participated in this effort, and the number of large dams in China increased dramatically, accompanied by the rapid growth of the number of relocated people. The government required individuals to sacrifice their well-being for the nation's development, and this approach was seen as legitimate rather than contradictory. Due to the collective ownership of all property and weak collective organizational structures, individuals, and communities did not present significant resistance. Relocation policies were often poorly designed and relocated people suffered from extremely low compensation and poor living conditions (Wang et al. 2013b).

4.3.1.1 Case Study 1: Sanmenxia Dam

Sanmenxia Dam was built on mainstream Yellow River to resolve sediment accumulation and flood problems (see Chap. 2 for the background information of the Yellow River and the Sanmenxia Dam). A fertile plain would be inundated by the reservoir and about 300,000 people needed to move. The new place the government chose for them was Qingtongxia in Ningxia Province, roughly 500 km upstream from their original location. The government used every means to convince the villagers to move. The government described Qingtongxia as the most productive and pristine region along the Yellow River, and the government promised villagers several times more farmland than in their original villages. Additionally, the government explained to the villagers that Sanmenxia Dam would protect hundreds of millions of people downstream from extreme floods, and what they were doing was to "move one household to save ten thousand households." The resettlement program was not smooth, but 300,000 people were moved to Qingtongxia before the impoundment of the reservoir (Tan and Liu 2003; Tianjin E-North netnews 2003). Sanmenxia Dam was completed in 1960, and due to poor design of the dam and the consequent sedimentation problems, more farmland was inundated 1 year later, and an additional half a million people were forced to move to Gansu Province hundreds of kilometers away (Tan and Liu 2003; Tianjin E-North netnews 2003).

When the villagers arrived at their new homes, they found that reality was very different from what was promised. The land was very rocky and precipitation was very low, resulting in extremely low agricultural productivity. It was almost impossible to make a living on the resettlement land, and most villagers fell into stark poverty. The "Sanmenxia refugees" became a symbol of people who sacrificed for the whole nation's well-being during the period 1949–1978 (Tan and Liu 2003).

4.3.2 *1978–1993: Opening Up*

"The reform and opening-up policy" implemented in 1978 was an historic turning point in modern China. The government admitted the mistakes of idealism and the political movements of the previous period. Central planning schemes were abandoned in some economic sectors, specifically agriculture and manufacturing. Accountability change during this transition was highlighted by devolution of power from the Central Government and the reduction of rigidity in administration. Individual farmers still did not own their farmland, but they had the right to use it, and they were recognized as the owners of the products from their farmland (Zhang 2011). Most industries were, however, still subject to central planning and the "market economy" was still restricted. The tragedies caused by improper compensation policies in previous periods made the government reconsidered its strong command and control strategy. Public officials became much more cautious in pursuing large-scale involuntary resettlement. New policies substantially increased the standards of compensation in terms of material wealth. Compensation for losses of embodied wealth and relational wealth was rarely considered (Wang et al. 2013b).

4.3.2.1 Case Study 2: Manwan Dam

Manwan Dam was one of eight cascading hydroelectric dams planned on the mainstream of the Lancang River (the Upper-Mekong River) to take advantage of its steep elevation drop. It was also the first dam in Yunnan Province, which took part in the "Transport western electricity to the east" strategy to meet massive energy demand in Guangdong, a much more developed coastal province. The project was completed in 1995 and had generating capacity of 1,250 MW, which is very small compared to other hydropower dams in China such as the Three Gorges Dam with installed capacity of 18,200 MW. The resettlement issues related to Manwan Dam are considered to be very typical of the period. More studies of the resettlement problems of Manwan Dam have been published than for any other dam of this scale (Yuan and He 2004).

According to official data, there were 3,208 people relocated in order to construct reservoirs. According to various independent studies, the actual number of displaced people was much higher, since landslides and other dam-related problems forced more people to move over time. After the dam was constructed, some local people still lived in the same villages, or just moved a short distance from their original villages, who were referred to as "near-relocated people." In some other villages, the impacts of the reservoir were so significant that the original villages were no longer usable and the communities had to be relocated to faraway places. These are the "far-relocated people" (Yuan and He 2004; Tilt et al. 2009).

When the near-relocated people moved, they were compensated according to area of their homes, and they were responsible for building new houses for themselves. They did not receive compensation for lost farmland, but the government cleared new land for them around the new village. The area of farmland per household decreased notably after moving, and the quality of the soil was worse than before. The lack of irrigation water also forced villagers to switch from growing rice to growing wheat or corn, which do not require intensive irrigation. The transportation condition for near-relocated villagers was very poor, which made access to healthcare and education difficult. Drinking water was also hard to get in new villages.

Compensation for far-relocated villages was better. Households were still responsible to build new houses themselves, but received fair compensation for old houses. The government was not able to clear enough farmland for all the villagers, so they provided them monetary compensation for shortfalls based on requirements of legislation. Agricultural incomes were insufficient at both locations, forcing villagers to go to cities to work in factories for several months a year. Since the new location was closer to the county town, transportation, healthcare, and education conditions were better than before. But some villagers noted that drinking water was insufficient at times. The relocation of displaced people into existing settlements put pressure on infrastructure, such as water supply, but investments were not made to upgrade infrastructure.

Compared to the Sanmenxia case, the government no longer required relocated people to sacrifice literally all their wealth for the nation's well-being in this period. Instead, to guarantee that relocated people get minimum compensation to maintain a living, the Central Government enacted legislation in 1991 to structure compensation practices for large hydro projects. According to this legislation, farmland was to be compensated at 3–4 times the average value of the yields over the last 3 years (The State Council of China 1991). Even though this was still a very low compensation standard, and one that would not cover losses incurred by local people even in terms of material wealth alone, it eradicated the extreme cases such as the "Sanmenxia refugees." Losses of embodied and relational wealth were not recognized, which made adaptation to the new environment difficult for relocated people.

4.3.3 1994–2000: State Meets the Market

China's transition from a centrally planned economy to a "socialistic market economy" was a gradual process. Different forms of market economic activities existed after 1978, but there were continuous debates on whether socialism and market economy contradicted each other. The legitimacy of market economy was not established until the end of 1993, when the central committee of the Party issued its watershed document, "The Decisions on Establishing Market Economy in China." This document removed most of the legal and ideological obstacles for the development of both state-owned and private companies. Accountability regime change in this transition was also gradual. With the progressive introduction of market institutions, the government no longer assigned arbitrary values in various transactions between different sectors. The market began to become the arbiter of price.

State-owned companies, rather than government agencies as in previous periods built most large dams during this period. The villagers also knew that companies, not the government, were taking their houses, farmland, and other assets. They no longer had the sense of "sacrificing (our) well-being for the whole nation's development," and they began to ask for more compensation. Compensation became a negotiation between residents, government, and state-owned companies (Wu 2004). However, due to lack of the information and ability to formally defend their own rights, residents were still at a disadvantage in negotiation. Little, if any, compensation was provided for losses of embodied and relational wealth (Wang et al. 2013b).

4.3.3.1 Case Study 3: The Three Gorges Dam

The Three Gorges Dam is not only the world's largest hydropower dam; it has generated the greatest number of relocated people. In 1985, it was estimated that the dam-formed reservoir would submerge the homes and environs of 726,000 people, and if population growth was considered, the number would reach 1,132,000 by 2008. Official documents have reported that after the completion of the relocation program in 2010, a total of 1,397,000 people had been relocated, including 190,000 in other provinces (Xinhua News Agency 2003b; Jackson and Sleigh 2000; Li et al. 2001).

Compensation in the case of Three Gorges Dam was largely based on the market price of the losses of relocated people. This approach differs markedly from setting a minimum compensation standard, as was the case in the Manwan Dam. In a document produced by the Central Government in August 1993, "The Three Gorges Project Development-oriented Resettlement Regulations," it was stated that compensation should be fair so that the relocated people could maintain or even improve their original standards of living. Further, the government committed to creating favorable conditions for the long-term development of displaced peoples' livelihoods. Another Central Government document in 1999 stated that the destination districts should provide favorable natural and economic conditions to the relocated people (Xinhua News Agency 2003b). Implementation of the compensation policy to "maintain or even improve original standards of living" involved evaluating the market prices of houses, land, and other assets, thereby improving the fairness of compensation in terms of material wealth.

In reality, relocated people were given houses and new-cleared farmland. The government also awarded monetary compensation at a rate many times more than their annual incomes. So, in term of material wealth the relocated people were well compensated. But the implementation of "developmental-oriented resettlement strategy" was unsuccessful. Very few training programs were provided to the relocated people, and after relocation a large percentage of people were unemployed. The government also built hospitals, schools, and other infrastructure projects in new built villages, but in most cases the investments could not meet the needs of the villagers (Xinhua News Agency 2003b). Therefore the losses of relocated people in terms of embodied and relational wealth were still substantially larger than the compensations they received.

4.3.4 2001–Present: Network Society

After 20 years of economic boom with an annual growth rate of 8 % on average, a large middle class was formed in China after 2000. These people were relatively well-educated, with stable incomes, and they had concerns about various social and environmental problems. Many environmental organizations were founded during this period, and they played active roles in efforts to protect biodiversity, reduce pollution, and mitigate environmental impacts of economic expansion. The widespread use of the Internet in this period also had far-reaching influences on China's society. Over a long history, state-controlled media had been the only information source for important events happened in China, and the government reported only the positive sides of its policies. But the Internet made it very difficult to hide information from the public. The rise of civil society and the inability of the government to control media drove an accountability regime change in which public engagement and participation enhanced transparency and openness in governance. Compensation policy design in this period was determined by the interaction among government, private sectors, and civil society, which made the negotiation and decision-making process much more complex (Wang et al. 2013b).

4.3.4.1 Case Study 4: Dams on the Nu River

The Nu River originates in the Qinghai-Tibetan Plateau, running through Yunnan Province in southwest China, and finally flows into the Indian Ocean. Debates about whether to build dams on the Nu River have gone on for 9 years since 2003, and the future of this unique river is still undetermined.

The possible exploitation of the Nu River for hydroelectricity was first incorporated into the national hydropower development plan in the 1980s. In 1999, the National Development and Reform Commission decided to start the procedure for constructing a series of dams on the Nu River. After public bidding, two institutes, the Beijing Investigation and Design Institute and the East China Investigation and Design Institute, were identified as being responsible for the design and planning of the cascade dam project. After 4 years of investigation and discussion, they presented a proposal to build 13 cascade dams with a total installed capacity of

21,320 MW, 1.2 times the capacity of the Three Gorges Dam. Until that time no one had doubted that the cascade dams would be built soon on this river (Dai 2004).

However, dramatic changes happened just 2 weeks later. Two environmental organizations, Green Watershed and Green Earth Volunteers, launched a campaign against the dam project on the Nu River that attracted public attention. Green Earth Volunteers convinced 62 influential people, including scientists, writers, journalists, and environmentalists, to sign a declaration to protect the natural status of the Nu River. Green Watershed conducted investigation on the standard of living of relocated people from Manwan Dam, and claimed that dam construction on the Nu River would have significantly adverse impacts on local residents. Many media services began to publish articles about the proposed dams, and most of them positioned themselves against the project. NGOs also organized international conferences, and even drew the attention of UNESCO, which recognized their petition and showed "concern" about the Nu River Dam project. Finally, then Chinese Premier WEN Jiabao called for the suspension of all the dam projects on the Nu River and requested a more detailed environmental and social impacts assessment (Cao and Zhang 2004).

In the years following this 2003 landmark action, proponents and opponents of dam construction on the Nu River have continued their seesaw battle. In 2005 a group of famous scholars who supported the project visited the region, and blamed the environmental organizations for being "biased" and "extremists." Some institutes also conducted the environmental impact assessments, which acknowledged some negative effects the project would have on the ecosystem and indigenous culture, but overall supported the dam project. Both sides have continued submitting petitions to the Central Government. The most recent petition was written by four geologists in February 2011, who warned that the Nu River runs along on a fragile geological zone, and that the dam project could trigger various geological disasters, including large earthquakes. Official agencies announced in March 2011 that the project was still under a "feasibility study" and that a decision could not be made until its completion (Lv 2011; Reuters Beijing 2011; Wang 2011).

After being hotly debated for 9 years, the future of the Nu River is still undecided. Regardless of the result, the debate has significant implications for China's dam construction history. It was the first time in modern China that civil society's opinions influenced the Central Government's decision-making process regarding infrastructure construction. It also represented a great leap in understanding the roles of different stakeholders in dam construction. The Nu River campaign signaled that the forces of civil society in China are emerging and growing. The expectation is that civil society will play more and more important roles in the future development of China's infrastructure and its natural resources policy and management.

The Nu River debate also improved the transparency in dam construction. Previously the public was only informed how much profit a dam project had generated, but citizens had never been told about the negative effects on environment or local people. But during the Nu River debate, all the positive and negative impacts of cascading dams were openly discussed via websites, newspapers, and TV shows. All stakeholders, including government officials, scientists, local people, NGOs, and hydroelectric companies, had opportunities to give their opinions and also consider opinions from other stakeholders. This deliberative environment enriched the knowledge of all stakeholders about dam impacts.

Since dam construction on the Nu River has not started, no people have been displaced. But the potential problems in resettlement have already been extensively discussed. One of the two NGOs that initiated the Nu River protection campaign, the Green Watershed, offered trips for local people on the Nu River to visit relocated villages associated with the construction of the Manwan Dam. After learning the adverse impacts of relocation on the villagers, the people along the Nu River became more skeptical of promises made by government and industry. Presumably, if the dams on the Nu River were to be constructed in the future, the government and electricity companies would have to invest much more in resettlement programs than in previous cases, particularly in embodied and relational wealth, as there is broader understanding of the implications of resettlement and the various dimensions of wealth lost by displaced people.

4.4 Government, Market, and Civil Society: New Coordination Mechanisms to Protect the Rights of Relocated People

Resettlement caused by dam construction is a complex problem that invites contemplation on what constitutes the public good and what ethical duties states and citizens have to individuals. By conceptualizing wealth as a multidimensional concept and by empirically tracing compensation policy over time, we are able to identify an historical pattern in the way displacement has been addressed within hydro projects in China. The pattern reflects commitments to compensate people more generously over time and an expanded definition of what compensation entails (i.e., cash payments for loss of property, skills training, investment in infrastructure). The pattern also reflects expansion over time in the number of actors with active roles in compensation policy. We observe interplay and complementarity among state, market, and civil society, the three pillars of governance. Examined through the specific lens of hydro project compensation policy, the three transitions in modern Chinese history highlight the significance of each of the three principle modes of sociotechnical coordination (Wang et al. 2013b).

In the first transition, the government learned from mistakes in the previous period and adjusted its resettlement and compensation practices. Since local minorities were no longer expected to sacrifice all of their assets to the national interests, compensation was expanded. However, without the engagement of market actors and civil society to complement and hold the state accountable, the rights of relocated people remained weakly recognized and weakly realized.

In the second transition, market mechanisms were introduced into compensation policy. The loss of material wealth began to be evaluated through reference to market prices, rather than fixed values the government assigned. On this basis compensation policy in material wealth was substantially improved. Relocated people were, however, still at a disadvantage in negotiations with the government and state-owned companies. Living in deep river valleys for generations, these people generally lacked the ability to formally defend their own rights. Cultural and organizational structures served to impede effective collective action, which further disadvantages local people.

The third transition, the information explosion that characterizes network society, introduced many new stakeholders into the resettlement process including NGOs, researchers, media, and international organizations. The widespread use of the Internet made aspects of the decision-making process transparent to all stakeholders, and society at large. With the support of organized and established civil society actors, relocated people gained more bargaining power in negotiations with commercial and state actors. Compensation came to include not only material wealth, but also embodied and relational wealth (Wang et al. 2013b).

The three transitions we identify in Chinese compensation policy illustrate the potential relevance of the paradigm shift "from government to governance" in the Chinese context (Stoker 1998; Rhodes 1996; Paavola 2007). Within this Western conception and analysis, admixtures of state, market, and civil society are seen as producers and implementers of policy, and they are seen as important elements of capacity to respond to social and environmental problems (Wang et al. 2013b).

Chinese governance continues to undergo profound transformations, and lessons from historical transitions could provide guides for policy making and reforms in the future. Our findings suggest that continuing and deepening interdependencies among public, private, and civil society actors point to more just treatment of people displaced by hydro projects and expanded capacity of local people to resist displacement. More generally, the trajectory in the previous transitions suggest that further expansion of transparency and public participation should be a focus in future reforms in order to enhance the legitimacy and accountability of government and to optimize the coordination mechanisms in governance (Wang et al. 2013b).

References

Cao H, Zhang P (2004) Civil society's force behind the suspension of Nu River Project. Economy (in Chinese)

Cernea M (1997) The risks and reconstruction model for resettling displaced populations. World Dev 25:1569–1587

Coleman JS (1994) Foundations of social theory. Belknap Press of Harvard University Press, Cambridge

Dai L (2004) The Nu River Dam debate: debate on what? Outlook Weekly (in Chinese)

Freudenburg WR (1986) Social impact assessment. Ann Rev Soc 12:451–478

Hollingsworth JR (2000) Doing institutional analysis: implications for the study of innovations. Rev Int Polit Econ 7:595–644

Holz CA (2008) China's economic growth 1978–2025: what we know today about China's economic growth tomorrow. World Dev 36:1665–1691

Jackson S, Sleigh A (2000) Resettlement for China's Three Gorges Dam: socio-economic impact and institutional tensions. Comm Post-Commun Stud 33:223–241

Li H, Waley P, Rees P (2001) Reservoir resettlement in China: past experience and the Three Gorges Dam. Geogr J 167:195–212

Lv M (2011) New Opponents against Nu River Hydropower. South China weekends. http://www.infzm.com/content/55856. Accessed September 5, 2012

Mazmanian DA, Kraft ME (2009) Toward sustainable communities: transition and transformations in environmental policy. MIT, Boston

Mulder MB, Bowles S, Hertz T, Bell A, Beise J, Clark G, Fazzio I, Gurven M, Hill K, Hooper PL, Irons W, Kaplan H, Leonetti D, Low B, Marlowe F, McElreath R, Naidu S, Nolin D, Piraino P, Quinlan R, Schniter E, Sear R, Shenk M, Smith EA, Von Rueden C, Wiessner P (2009) Intergenerational wealth transmission and the dynamics of inequality in small-scale societies. Science 326:682–688

Paavola J (2007) Institutions and environmental governance: a reconceptualization. Ecol Econ 63:93–103

Reuters Beijing (2011) Nu River plan is still under feasibility analysis. http://cn.reuters.com/article/macroeconomicsNews/idCNnCN165548920110308. Accessed May 15, 2012

Rhodes RAW (1996) The new governance: governing without government. Political Studies 44: 652–667

Sadler B, Verocai I, Vanclay F (2000) Environmental and social impact assessment for large scale dams, WCD Thematic Review V.2 prepared as an input to the World Commission on Dams. WCD, Cape Town

Stoker G (1998) Governance as theory: five propositions. Int Soc Sci J 50:17–28

Tan Y, Liu H (2003) Sanmenxia Dam: the dangerous status of the former "first dam in China." http://www.people.com.cn/GB/huanbao/1072/2206736.html. Accessed May 15, 2012

Tang C (2009) Policies and practices in resettlement caused by dam construction in China. http://www.chinawater.net.cn/Journal/slxb/20009z/14.html

The State Council of China (1991) Statute of land compensation and people relocation in large hydraulic project construction

Tianjin E-North Netnews (2003) A 50 year reflection on The Sanmenxia Dam. http://news.enorth.com.cn/system/2003/12/03/000678411.shtml. Accessed May 15, 2012

Tilt B, Braun Y, He D (2009) Social impacts of large dam projects: a comparison of international case studies and implications for best practice. J Environ Manage 90:S249–S257

Vanclay F (2003) International principles for social impact assessment. Impact Assess Project Appr 21:5–11

Wang Z (2011) Wang Shucheng: no rush on nu river hydropower development. China Economic Weekly

Wang C, Li A, Liu D (2007a) Management analysis of resettlement caused by dam construction in China and in the world. Expr Water Res Hydropower Inf 28:27–30

Wang P, Lassoie JP, Dong S, Morreale SJ (2013a) A framework for social impact analysis of large dams: a case study of cascading dams on the Upper-Mekong River, China. J Environ Manage 117:131–140

Wang P, Wolf SA, Lassoie JP, Dong S (2013b) Compensation policy for displacement caused by dam construction in China: an institutional analysis. Geoforum 48:1–9

WCD (2000) Dams and development: a new framework for decision-making. World Commission on Dams

Wu F (2004) Residential relocation under market-oriented redevelopment: the process and outcomes in urban China. Geoforum 35:453–470

Xinhua News Agency (2003b) Background of the Three Gorges Project. http://news.xinhuanet.com/ziliao/2003-05/30/content_894678.htm. Accessed May 15, 2012

Yuan X, He D (2004) Land cover remote sensing monitoring on Manwan Dam immigrant settlement. J Mt Sci 22:267–271

Zhang L (2011) The political economy of informal settlements in post-socialist China: the case of chengzhongcun(s). Geoforum 42:473–483

Chapter 5
Conclusions: The Future of Large Dams in China

5.1 Trends and Debates of Large Dams in China

This book included a review of large dam construction worldwide and at China's national level. While the worldwide review mostly discussed the general characteristics of dam construction, the Chinese review focused on the particular features of large dam construction in China.

The worldwide review first discussed the history, distribution, and the trends of large dams. Even though the history of dams is as long as that of human civilization, the overwhelming majority of dams were built after 1900, with the peak in construction occurring between 1970 and 1975. Five regions that have the highest dam density are East Asia, South Asia, North America, Europe, and Southern Africa. The distribution of dams is uneven between developed and developing countries: the former have already developed more than 70 % of their economically feasible dam potentials, while the latter have only developed about 30 %. Hence, most of the dams currently under construction or being planned are in developing countries, and therefore, the trend of future dam construction worldwide mainly depends on the pace of economic development in these countries.

The major purposes for dam construction include irrigation, water supply, flood control, electricity generation, and navigation. Some of the objectives are well achieved, which are the cases for most hydropower dams. But many dams intended for irrigation or water supply have fallen short in meeting their original goals. Dams for flood control could be a double-edged sword: they could regulate natural flow and reduce the magnitudes of floods, but also could cause unintended consequences, such as sedimentation of riverbeds, and making flood disasters even more pronounced. Some dams are built to raise the water level to improve navigation conditions, but this may come at considerable environmental and social costs.

General environmental impacts of large dams include inundating terrestrial ecosystems, altering natural river flow regimes, blocking routes for migratory species, increasing sedimentation, and enhancing greenhouse gas emissions. Most of the changes to natural systems are irreversible; therefore, it is necessary to conduct reasonable

P. Wang et al., *The Large Dam Dilemma: An Exploration of the Impacts of Hydro Projects on People and the Environment in China*, DOI 10.1007/978-94-007-7630-2_5,

environmental impacts assessment before the construction process is approved and to identify proper mitigation measures that should be applied in the project.

There are different types of social impacts of large dams that occur at different levels, including those on agriculture, local livelihood, social inequity, and culture. Compared with environmental impacts, social impacts are harder to identify and assess, and therefore are easier to be overlooked during the decision-making process. However, a comprehensive understanding of social impacts is essential to the fair distribution of benefits and costs over all stakeholders. Therefore, incorporating a social impact assessment into the decision-making process is as important as applying an environmental impact assessment before dams are constructed.

Debates around large dams generally fall into three categories: economic feasibility, environmental sustainability, and the social equity issues. Large infrastructure projects like dams provide direct and indirect economic benefits, and have been used for a long time as an important means to stimulate economic development. But there are also externalized costs of dams, which make the benefit-cost analysis problematic. The environmental impacts of dams are difficult to deny, but the debates are in most cases on whether opponents of dams exaggerate those impacts. Social equity debates focus on one key issue: whether the groups who suffer the most from dams also benefit the most, or get fairly compensated for their losses.

In China there were very few large dams before 1950. However, during the second half of twentieth century China built about half of all the large dams constructed worldwide. Most dams built between 1950 and 1966 were aimed at irrigation and managing hazardous rivers. In contrast, most dams built after about 1980 primarily focused on generating electricity, due to the rapid increase in energy demand as China modernized. Currently, electricity generation is the single most important driver for dam construction in China. The Ministry of Water Resources identified 13 hydropower bases in different river basins in China, which cover almost all the river reaches that have technically feasible hydropower potential. The early construction projects did not account much for social impacts, but there has been a steady progression of policy and inclusion of stakeholders ever since.

The case studies in this book represented the most important or controversial dam projects in China. The Three Gorges Dam is by far the largest hydraulic project in the world, and it was finally build after a 70-year-long debate. With multiple functions to control floods, improve navigation, and generate electricity, it has also caused substantial environmental and social impacts in the reservoir-affected area. The Xiaolangdi Key Water Control Project is the most critical facility on the Yellow River to manage flow and mitigate sedimentation. Since the Yellow River is a seriously degraded river system, Xiaolangdi Dam is generally successful in harnessing this unpredictable river. The cascading dams on three rivers in Southwest China, the Nu, Lancang, and Jinsha were discussed last. Dams on these rivers are all intended for electricity generation, and use the same model of cascading development, which means planning a series of dams on one river and using the profits from earlier complete dams to finance the building of the rest. This cascading model could be an economically efficient means for developing entire river systems, but the social and environmental impacts are also extended to entire watersheds.

5.2 Understanding the Environmental Impacts on Biological Diversity and Ecological Integrity

Environmental impacts are among the most significant concerns of large dam construction. This book reviewed the environmental impacts of large dams from the perspective of biological diversity and ecological integrity. In Chap. 3, we showed generally the environmental impacts of large dams in major river basins throughout China. After the general review of the environmental impacts, we also introduced frameworks for assessing these impacts at watershed and ecosystem scales based on the case studies in the Upper-Mekong river basin.

Inundation of the reservoir area is the most obvious direct impact a dam will have on the environment. Most large dams are located in valleys, which often have relatively high biodiversity. Thus, dam construction inundates critical habitats for many plant and aquatic species. Reservoir inundation could also cause a change in land use patterns. After impoundment, local people may be forced to move up the hillsides and clear new farmland where forests may have been before the construction of the dam. This change in land use also might aggravate soil erosion, further threatening biodiversity. In Chap. 3, we developed the vegetation impact index (VII) and the ecological risk grades to quantify the impacts of the large dams on plant diversity.

Many aquatic species in rivers have specific migratory patterns during their life cycles. During different life stages, such as spawning, juveniles, and sexual maturation, many species shift to different reaches of a river. Dams can block migratory paths, and reduce the populations of many species, possibly driving some species to extinction. Some large dams have established fish bypasses (ladders) as a means for mitigating the negative effects on migratory species, but the percentage of large dams with such structures is very low, and ladders only aid the movement of certain species. Also, these structures sometimes do not work effectively, since the migration of different species requires various navigational cues, such as strong currents.

Dam construction is considered one of the most significant alternations to natural hydrologic systems. The evolution of a river ecosystem is a process of adapting to the flow regime over several thousands of years. However, dam construction causes rapid changes in natural flow regimes in several aspects. The flow regime changes, that in associated with water quality alternation, will significantly impact the integrity of river ecosystems. In Chap. 3, we structured the fish index of biological integrity (F-IBI) and the planktonic index of biological integrity (P-IBI) to quantify the impacts of large dam on the ecological integrity of river ecosystem.

Another alteration of dam projects to the natural system is sedimentation, which is not only a technical problem for the dam itself, but also has noticeable environmental consequences. For many large dams, serious sedimentation shortens the service life and long-term sustainability of the facility. Dams capturing sediment moving downstream can also cause the erosion of downstream channels and tributary headcutting. Estuaries and deltas need sediments from upstream to offset the erosion by waves and tides. Dam construction not only captures the sediment, but also reduces the amount of water flowing downstream because of evaporation and water consumption by agriculture and/or industry. Therefore, seawater might encroach and erode estuary areas.

Hydropower has been considered a clean energy source with lower greenhouse gas emission rates than electricity generated by fossil fuels. However, studies show that decomposing organic matter in reservoirs emits a large amount of greenhouse gases. This problem is more severe in shallow reservoirs in tropical regions. However, since natural lakes and wetlands also emit greenhouse gases, whether the contribution of reservoirs to global climate change is significant remains controversial.

5.3 Understanding the Social Impacts and Compensation Policy

Large dam projects will not only dramatically change the physical conditions of the rivers, but also will have significant impacts on local communities. In Chap. 4 we identified three classes of wealth for the affected people, namely material wealth, embodied wealth, and relational wealth, and comprehensively compared the loss and compensation in each dimension of wealth. Case studies showed that the division of three dimensions of wealth can help researchers and policy makers better understand the multilevel social impacts on relocated people. Dam construction has direct impacts on material wealth, primarily by submerging houses and farmlands. Material wealth is easy to understand, and therefore its loss is most often equitably compensated, or even overcompensated. But embodied wealth and relational wealth are generally affected in indirect manners, and thus they are more difficult to recognize by policy makers. Hence, most unfair compensation practices arise from ignoring either embodied wealth or relational wealth. However, the impact is not always negative, as some dam projects bring concurrent opportunities for villagers to enhance these two types of wealth, such as by providing more job opportunities or improved infrastructures.

Recommendations for more reasonable compensation policies were also identified based on the analysis of the three dimensions of wealth. In Chap. 4 we showed that it is critical to systematically assess the three dimensions of wealth loss caused by dam construction, and that in order to improve the accuracy of the assessment, a transparent decision-making process is necessary. Then, to achieve the goal of fair compensation, the government needs to compare which dimensions of wealth loss are covered and fairly compensated by the policy and compare them with the actual wealth loss in each dimension as perceived by villagers. The ideal compensation policy should compensate fairly in all three dimensions. The compensation policy should also take social inequity effects into consideration and make sure to avoid uneven impacts on different individuals or communities that would widen the inequity gap between them.

5.4 New Decision-Making Schemes in Large Dam Projects

Large dam construction is not merely an engineering problem, but is rather at the center of resource exploitation, environmental protection, and social development. The complexity of this problem demands that resolving it will require interdisciplinary knowledge and collaboration. Figure 5.1 illustrates the interactions among the three components of large dam construction. Owing to the inter-relationships between these components, failing to address any one of them likely will result in the eventual failure of the project. While it is critical to coordinate the demands arising from all three components, it is also important that experts in different fields study each component in detail. The following suggestions are provided for improving the understanding of each component.

Economic feasibility is at the center of resource exploitation. When planning to build a dam, resource potential, technological capacity, finance, and market are all among the factors that need to be evaluated. Beside these considerations, it will be helpful to research the financial and physical performances of dams, such as the possibilities for overrunning estimated budgets and/or failing to meet planned objectives. The World Commission on Dams' (2000) report did interesting statistical analyses of the physical and financial performances of dams around the world. However, it might be more persuasive to make thorough inquiries into the reasons behind the high or low performances of dams with different purposes, such as why most of irrigation dams fail to meet their original goals, while most hydropower dams not only recover their costs, but also surpass their original objectives. Another possible research topic involving economic feasibility would be to compare the profit rate of large dams with other alternatives. For example, comparing hydropower plants with solar or wind power plants, or comparing the economic effects of irrigation dams with new, innovative water saving technologies. Such comparisons will help developing economies decide if they should build large dams or pursue other options.

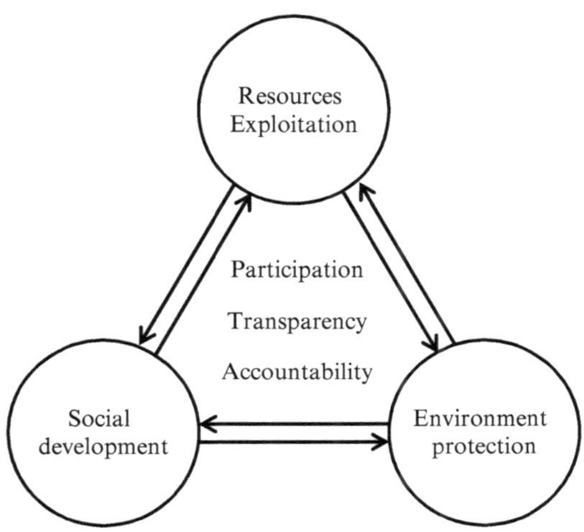

Fig. 5.1 Economic feasibility, environmental sustainability, and social equity

Environmental sustainability is another goal that large dam projects should strive to achieve. Much research has focused on the environmental impacts of large dams, including studies on immediate changes in terrestrial and aquatic ecosystems, geological features, and water quality following the construction of large dams. Future research on environmental sustainability problems should also focus on long-term impacts, such as greenhouse gas emission rates, regional climate changes, and relationships between reservoir size and the frequency of earthquakes. Since a substantial percentage of freshwater on the planet is held in large dam reservoirs, these long-term impacts to the global environmental sustainability should not be neglected.

Social equity was not emphasized in dam projects until the recent decade. Even though large dams have been used as an important means for social development, the distribution of benefits and costs are not equitable in most cases. The following suggestions are potential directions that future research on the perspective of social equity should advance. First, this book proposed a framework for better understanding the social impacts of large dams, where wealth was divided into three categories. Future researchers may work on designing a more scientific and reasonable division of wealth. For example, the definition and examples of embodied and relational wealth need to be further specified. Second, a global decision-making framework is necessary to fully understand the social impacts, and the World Commission on Dams has made great contributions in building such a framework. However, large dam problems vary significantly between different regions, therefore geographic and socioeconomic differences should be taken into consideration and necessary adjustments should be made to the framework to better fit specific situations. At last, certain universal standards need to be developed to ensure basic social equity during and after the dam building process, in order to make sure that all the stakeholders are fully informed, the decision-making process are participatory, and the policy designs are transparent.

The future of large dams in China is shaped, and being shaped, by the different interests held by different stakeholders, and the complex relationships between economic, social, and environmental systems. The solution to the large dam dilemma should not be a simple judgment of building dams or not. It should be a mechanism that ensures that different factors in human and nature systems are considered, and interests of different groups are respected and protected. Participation by all key stakeholders, transparency in the decision-making process, and accountability in the implementation of new projects comprise the cornerstone for this mechanism, which assures that the decision-making process is scientific-based, efficient, and fair. In sum, this reasonable decision-making mechanism, rather than a moral or economic judgment, is the answer to the large dam dilemma.

Reference

WCD (2000) Dams and development: a new framework for decision-making. World Commission on Dams

About the Authors

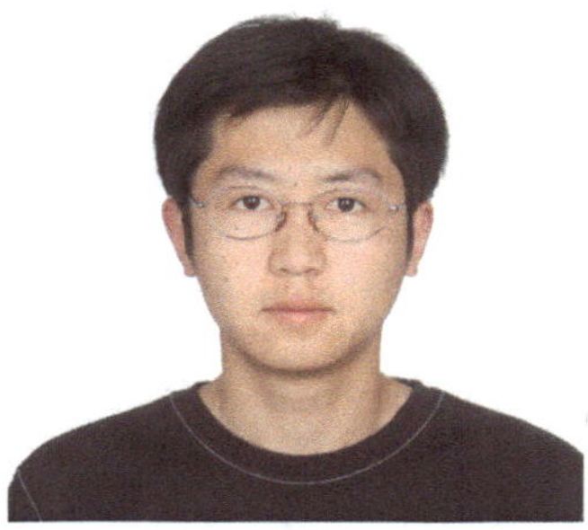

Pu Wang is a Ph.D. candidate at Department of Natural Resources, Cornell University, USA. His Master's research focused on the socioeconomic impacts analysis of large hydro-projects. His current research involves market-based instruments for natural resources management, particularly the economics and institutional analysis of payment for ecosystem services (PES) schemes. Email: pw294@cornell.edu. Mailing address: Fernow Hall, Room 111, Cornell University, Ithaca, NY 14853, USA.

P. Wang et al., *The Large Dam Dilemma: An Exploration of the Impacts of Hydro Projects on People and the Environment in China*, DOI 10.1007/978-94-007-7630-2,

Shikui Dong is a full Professor at School of Environment, Beijing Normal University and an Adjunct Professor at Department of Natural Resources, Cornell University. He also serves as Vice Secretary for Chinese Grassland Society and an Invited Panel on Environment Impact Assessment Center for China's Ministry of Environment. He has authored over 150 publications, about 30 symposium proceedings and abstracts, and seven book chapters in recent years on ecological restoration, natural resources management, as well as strategies for sustainable development. Email: dongshikui@ sina.com. Mailing Address: School of Environment, Beijing Normal University, No. 19, Xingjiekou Waidajie, Haidian District, Beijing, 100875, P.R. China.

James P. Lassoie is a Professor in the Department of Natural Resources and an International Professor of Conservation in the College of Agriculture and Life Sciences at Cornell University. Originally trained as a forest ecologist at the University of Washington in Seattle, he now focuses on community-based conservation science and management and has worked extensively in Africa, Asia, Latin America, and North America. He holds an Adjunct Professor appointment in the School of Environment at Beijing Normal University and has worked in China since 1999. Professor Lassoie has over 150 scholarly research publications and is currently directing the development of a web-based educational system, www.conservation-bridge.org. Email: jpl4@cornell.edu. Mailing address: Fernow Hall, Room 201, Cornell University, Ithaca, NY 14853, USA.

Acknowledgements

In the process of completing this book, we got support from many people and institutions. We would like to express our appreciation to all those who have given us support.

We are deeply grateful to Dr. Ruth Sherman at Cornell University, who provided very valuable suggestions on the contents and language of this book and also did excellent editing work for the manuscript.

We also thank Dr. Steven Wolf and Dr. Stephen Morreale at Cornell University, and Li Jinpeng, Zhao Qinghe, Li Xiaoyan, Liu Shiliang, Peng Mingchun, Yang Zhifeng, Zhao Chen, Deng Li, Yang Yuejie, Wang Cong, and Zhang Zhaoling at Beijing Normal University for their contributions to the papers that we coauthored and cited in the book.

We also need to express our appreciation to many people who helped us in the field research. Xiao Ziwei at Rochester University and Li Jinpeng, Mei Ying, Dai Jing, and Chen Shaoqing at Beijing Normal University helped us conduct field work in Yunnan. Dr. Liu Ende at the Kunming Institute of Botany, Chinese Academy of Science helped us in our data collection processes.

This research received financial supports from Cornell University, Key Program of the National Natural Foundation of China, key project (Grant No. 50939001), and Nonprofit Environment Protection Specific Project of China (Grant No. 201209029-4).

P. Wang et al., *The Large Dam Dilemma: An Exploration of the Impacts of Hydro Projects on People and the Environment in China*, DOI 10.1007/978-94-007-7630-2,

MIX
Papier aus verantwortungsvollen Quellen
Paper from responsible sources
FSC® C105338

If you have any concerns about our products,
you can contact us on
ProductSafety@springernature.com

In case Publisher is established outside the EU,
the EU authorized representative is:
Springer Nature Customer Service Center GmbH
Europaplatz 3, 69115 Heidelberg, Germany

Printed by Libri Plureos GmbH
in Hamburg, Germany